ADVANCE PRAISE FOR
THE DNA OF STRATEGY EXECUTION: NEXT GENERATION PROJECT MANAGEMENT AND PMO

"In a world where there are more questions than answers every leader will need to learn to dance to a different beat. In this insightful book, Jack Duggal has cracked the DNA of Strategy Execution. Ignore these insights at your own peril."

—Dr. Tony O'Driscoll
Global Head, DukeCE Labs, Duke Corporate Education
Fuqua School of Business, Duke University

"With the customer at its core, the DNA of Strategy Execution is a fascinating examination of the strands that connect strategy and execution to lasting change. A must have read for strategy and project professionals looking for the agility to thrive on the edge of chaos."

—Ellen Kelly
Associate Director for Strategy and Planning
Research Foundation for SUNY

"A brilliant approach to organizational project management essential for an organization to execute on its business strategy in today's dynamic environment. A must read for all leaders, managers, and executives in start-ups or traditional businesses."

—Devesh Dayal
Global Head of Business Intelligence
Head of US Asset Management Technology
Deutsche Asset Management
Deutsche Bank

"The DNA of Strategy Execution takes an in-depth look at today's complex, challenging, and competitive business environments. We can no longer function the old way. This book inspires you to think differently and find a new approach in bringing corporate strategies and execution closer together."

—Brenda Horton
President, PMI Westchester
Senior Project Manager
IBM T.J. Watson Research Center

"The DNA of Strategy Execution really encourages and inspires the reader to take a step back and look at current processes from a new perspective. It is highly applicable as a tool to create or transform project management and PMO practices in today's world, particularly in product and software development."

—Amanda Adams
Director of Project Management
MINDBODY Online inc.

"The book is so well written and includes multiple angles and views on executing strategies in the world of today and the world of tomorrow. It's a must-read book and a must-implement book for all of us who want to take the steps into the future with focus on strategy execution and understanding how to optimize the value of it."

—Stig Villadsen
Strategy-Execution Advisor
Denmark

THE DNA OF
STRATEGY EXECUTION

THE DNA OF STRATEGY EXECUTION

Next-Generation Project Management and PMO

Jack Duggal

WILEY

Library of Congress Cataloging-in-Publication Data:

Names: Duggal, Jack, author.
Title: The DNA of strategy execution : next generation
   project management and PMO / Jack S. Duggal.
Description: Hoboken : Wiley, 2018. | Includes bibliographical references and
   index. |
Identifiers: LCCN 2017049057 (print) | LCCN 2018000078 (ebook) | ISBN
   9781119378891 (pdf) | ISBN 9781119378730 (epub) | ISBN 9781119278016
   (hardback)
Subjects: LCSH: Project management. | Strategic planning. | BISAC: TECHNOLOGY
   & ENGINEERING / Industrial Engineering.
Classification: LCC HD69.P75 (ebook) | LCC HD69.P75 D8364 2018 (print) | DDC
   658.4/012—dc23
LC record available at https://lccn.loc.gov/2017049057

Cover Design: Wiley
Cover Image: © Jezperklauzen/Getty Images

Printed in the United States of America

SKY10059696_111023

To my parents and Nina, with much love and appreciation ...

CONTENTS

6 GOVERNANCE 121

THE DNA OF
STRATEGY EXECUTION

1

INTRODUCTION: STRATEGY EXECUTION IN A DANCE-WORLD

"The real voyage of discovery consists, not in seeking new landscapes, but in having new eyes."

Marcel Proust

Imagine your organization is a mirror. In the beginning, when you are getting started, you can look in the mirror and see; you can connect and organize various aspects of the business, and it is manageable. As the organization grows, the mirror multiplies. There is addition of new mirrors of different sizes and shapes. Soon, it starts to get muddled with multiple mirrors and, you can't see clearly anymore. It appears that everyone seems to be working hard, but there is confusion. Execution does not mirror strategy, and there is no alignment between different parts. It makes sense to organize the mirror into more manageable divisions. You reorganize and break down the mirror with good intentions to better manage it. What do you think predictably happens?

On the surface, it looks like the resulting divisions and silos are well organized and working. But you don't realize that the organization is like a mirror, and as you divide and reorganize it, it is bound to crack and fracture in unpredictable ways. It is a broken mirror, and no matter how hard you try to bring it together, it is not going to look the same again. It causes organizational dysfunction, finger pointing, mixed priorities, and confusion—far from the much-needed clarity and coherence in an increasingly complex world.

Does this sound familiar? This is far too pervasive in organizations, from start-ups to incumbent companies and government organizations only with exponential complexity.

A 100-year-old iconic company that has been a leader in its domain has enjoyed sustained growth and profitability. Over the years, they have excelled at fine-tuning supply chains and gaining efficiencies. But like many businesses, it is struggling to survive and reinvent itself as it faces declining revenues and disruption from digitization and related competition. Recently, it reorganized into three business units to become more customer focused. There are over 100 projects and initiatives across these business units, and some of them are redundant. Each of the business heads is busy reorganizing and ramping up their areas. What is a predictable outcome? Is this going to help bring the organization together and get focused to meet the challenge head-on? What if they put together an organization like a project management office (PMO) to review and prioritize projects at the enterprise level? The challenge is that they have had several PMOs over the years with marginal success. Some were perceived as the process police and others as bureaucratic overhead. How can the PMO reinvent itself and support strategy execution in a holistic way?

On the other end of the spectrum is the typically chaotic world of start-ups. Too many priorities that keep changing. Disdain for process and governance. Decisions are made, but no one knows who made them or who is in charge. Products and offerings keep changing. Frequent reshuffling and reorganizing. It seems like people are working hard and putting in a lot of hours, but there is a disconnect between strategy and execution. Culture is pervasive with heroism and firefighting as they espouse flat structures and holocracy. Too much time and money are burned in the name of failing fast, speed and agility. How could we maintain the start-up spirit of innovation and agility with some structure and discipline?

"Look out the window, not in the mirror," challenged legendary management consultant, educator, and author, Peter Drucker. As we look out the window and reflect the different organizations we have worked with over the past 17 years, crucial questions resound consistently: How can we connect the siloes and link the various execution activities with the overall strategy in a holistic way? How do we manage in a disruptive world that is constantly in flux? Are we stuck in traditional management approaches that don't work anymore? We are trying iterative, incremental, and agile approaches, but do not see the results—what do we do?

In the quest for better organization and management, many practices have continued to evolve over the past 150 years—from Fredrick Taylor's scientific principles of management to Six Sigma, to current-day lean, agile, and hordes of variations in between. These practices have been applied with varying degrees of success depending on the organizational context, culture, and time frame. Some emphasize operational efficiencies and zone-in on execution; others focus on strategy and business effectiveness, while others espouse governance and control. In the nineteenth century, management consisted of six functions, according to Henri Fayol, considered to be one of the founders of modern management: forecasting, planning, organizing, commanding, coordinating, and controlling. Toward the end of the twentieth century, business management came to consist of six separate branches: financial management, human resources, information technology, marketing, operations, and strategic management.

As organizations evolve in size, scale, and complexity, there is more and more separation between various elements. In particular, the chasm between strategy and execution grows to the point that it is hard for the people involved in executing projects to be able to connect to the strategy or be aware of the business purpose and alignment, while the people responsible for strategy blame execution for not achieving results and vice versa.

The questions that sparked this book are: What if we could decode the DNA of effective strategy execution? Just as DNA contains the genetic instructions used in the development and functioning of all known living organisms, is there a code or blueprint containing the elements of management and strategy execution? How could the strands of the DNA be used to connect strategy and execution in a holistic way? How can we develop the next generation of practical project management and PMOs practices that can complement and enhance contemporary approaches like lean and agile? Is there a way to design an organizational mosaic and connect disparate pieces that provide sustainable results and value holistically?

This book is designed to inquire into these questions to gain new perspectives on age-old management challenges and illuminate better ways to organize, manage, and execute strategy.

The future of work is project-based, and more and more organizations are project-based. In this book we focus on how organizations can identify and develop the DNA elements to build a strategy-execution platform with next-generation project management and PMO capabilities. As we will demonstrate, decoding the DNA of strategy execution can be beneficial in any organization—whether it is an incumbent company, or a startup, or a non–project-based operational environment, the DNA of management is the same and can be applied as a foundational aspect alongside other approaches.

Many organizations have implemented varying degrees of project management practices and established PMOs to coordinate and monitor projects and strategic initiatives. PMOs have become a common fixture in organizations but are not necessarily perceived as a high value, breakthrough management idea. They are typically implemented in a limited way based on traditional management paradigms. According to multiple surveys, including an ongoing survey we have been conducting at the Projectize Group since 2005, the success rate has hovered in the 50 percent range over the past 12 years.

RECOGNIZE THE DANCE: THE NEED TO RETHINK AND REDESIGN

"Any company designed for success in the 20th century is doomed to failure in the 21st."

David S. Rose, Angel Investing

How do we redesign project management and PMOs for the twenty-first century? First, we have to start by understanding the foundation from which traditional management processes have evolved. As the world was becoming more industrialized in the twentieth century, there was a need for systematic approaches and standardization to gain efficiencies. These scientific principles of management popularized by Fredrick Taylor are based on a mechanical mindset—a factory model where work is done, or projects are completed, using structured processes in a controlled environment that can deliver predictable and consistent outputs each time. The processes, tools, and related techniques are based on a deterministic and reductionist approach. They are based on linear cause-and-effect thinking. The organization is viewed as an inanimate decision-making machine that works with process and technology designed to scale efficiency and take the human out of the process. It relies on manuals, step-by-step instructions, and detailed specifications. You determine the scope and decompose it into a work breakdown structure, which is also the core of project management techniques. This mechanical view has prevailed for most of the twentieth century, and for the most part, you can argue it has served us well. This approach has served us well, but for deterministic projects, where the problem and solution definition is known or knowable. But what do you do if the solution and related scope is unknown or unknowable, and requires learning over a period of time??

The reality of the world and today's business environment can be best characterized by the acronym **DANCE: D**ynamic and changing, **A**mbiguous and uncertain; **N**onlinear, **C**omplex, and **E**mergent and unpredictable—dynamic and constant change, driven by disruptive factors and shifting stakeholder needs and priorities. There is ambiguity and uncertainty, the situation is ambiguous or not clear and can be interpreted in different ways, and it is uncertain which way it will go. The direction is not clear, and there is a lot of uncertainty about the future. Unlike stable environments, where things are linear and expected, in a nonlinear world it is hard to ascertain the cause and effect, the output is not proportional to the change in the input, and it is therefore, hard to plan or manage the unexpected. The environment is complex because of the multiplicity of stakeholders involved, the number of interactions, and the sheer number of linkages and dependencies. It is not clear who all the stakeholders are, and the identified stakeholders are indecisive—they do not know what they want. Scope, requirements, solutions, and stakeholders are emergent and unpredictable and it can be hard to plan top-down in a continually shifting landscape.

DANCE is similar to VUCA, an acronym used to describe the volatility, uncertainty, complexity, and ambiguity, initially used in the military in the 1990s. DANCE is broader and better reflects today's business environment. It emphasizes the nonlinear and emergent aspects, which are not called out in VUCA.

To deal with the DANCE you need a fundamental shift to an organic mindset—a deep understanding and appreciation that today's organizations are nonlinear and complex due to the multiplicity of activities and dependencies, and unpredictable due to the intricacies and overlaps among people, information, and connections.

The conventional approach to implement project management or PMO is based on the mechanical view and deterministic methods of classic project management with a heavy emphasis on the linear scope, plan, execute, and control (SPEC) processes. But reductionist plans based on sequential tasks and dependencies do not seem to hold in a nonlinear, changing, and unpredictable project reality. For example, in a project, if you have one of the aspects of a project that is not well defined and is ambiguous and has three possible outcomes, it may be relatively easy to manage. If you have two areas of ambiguity with four possible outcomes, now you have 16 possibilities. As the areas of ambiguity increase, the possible outcomes grow exponentially—five areas of ambiguity with four possible outcomes each present 625 options. How do you plan to accommodate and deal with this many options? The challenge gets further compounded when the path forward is not clear, and you combine it with the sheer number of uncertainties surrounding the project.

The more independent steps that are involved, the more opportunities for failure. Let's say a start-up is working on a new product offering that has six steps. Each step has an 80 percent probability of success, which means 2 out of 10 times a step may not succeed. Since each step is independent the probability multiplies, and now the chance of going live with this product is only 26 percent, meaning they should expect success one time out of four. Even though it looks like an 80 percent chance of success in the beginning, each time we add a step, the probability of success reduces. It is easy to overestimate the probability of success based on the uncertainty of a number of steps, people, and decisions involved in a DANCE-world.

Which Game Are You Playing?

It is also important to distinguish the game you are playing. If you are flying an airplane, you rely on checklists and follow air safety rules and established processes. Aviation is a great example of safety and reliability based on established procedures, documentation, and learning from failures. Air travel remains the safest in terms of deaths per passenger mile. However, we must recognize that flying an airplane is different from designing and building an airplane and requires a different approach, as evidenced by the challenging development of the Boeing 787 Dreamliner project, which was $12 billion to $18 billion over budget and many years behind schedule. Projects like Boeing's Dreamliner are examples of the DANCE in action—the complete solution is not known at the start, there are many options, redesign of not just the structure but also the materials (use of composite materials), new and untried technology, complicated supply chains, and multitier outsourcing. See Figures 1.1 and 1.2, Building the plane versus Flying the plane.

It is important to understand the distinction between designing and building versus flying the plane as you implement project management practices. Do you recognize the need for a different approach?

It is like playing two different games—pool and pinball (Figures 1.3 and 1.4, Playing Pool versus Playing Pinball).

Pool (billiards) is based on linear cause and effect. Mastery requires understanding the rules and dynamics of the game, and with deliberate practice, you can control and master the variables and expect predictable outcomes. Pinball, however, has all the DANCE characteristics—as soon as the ball is launched, you do not know where exactly it is going to go. It may hit a target and gain momentum or slow down; it is dynamic and unpredictable. To succeed at pinball, you need to recognize that the nature of the game is different and requires a different approach. You can't follow a solid plan or keep flipping aimlessly. The irony is that organizations do not recognize

Selected component and system suppliers.

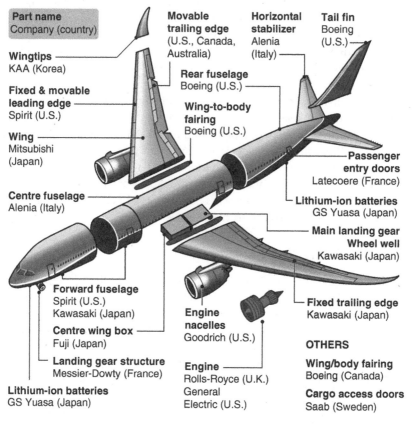

Part name
Company (country)

Wingtips
KAA (Korea)

Fixed & movable leading edge
Spirit (U.S.)

Wing
Mitsubishi
(Japan)

Centre fuselage
Alenia (Italy)

Movable trailing edge
(U.S., Canada, Australia)

Rear fuselage
Boeing (U.S.)

Wing-to-body fairing
Boeing (U.S.)

Horizontal stabilizer
Alenia
(Italy)

Tail fin
Boeing
(U.S.)

Passenger entry doors
Latecoere (France)

Lithium-ion batteries
GS Yuasa (Japan)

Main landing gear Wheel well
Kawasaki (Japan)

Forward fuselage
Spirit (U.S.)
Kawasaki (Japan)

Centre wing box
Fuji (Japan)

Landing gear structure
Messier-Dowty (France)

Lithium-ion batteries
GS Yuasa (Japan)

Engine nacelles
Goodrich (U.S.)

Engine
Rolls-Royce (U.K.)
General
Electric (U.S.)

Fixed trailing edge
Kawasaki (Japan)

OTHERS

Wing/body fairing
Boeing (Canada)

Cargo access doors
Saab (Sweden)

Figure 1.1: Building the Plane
Source: APG 060 – The Pilot pope – Airline Pilot Guy.

Figure 1.2: Flying the Plane
Source: José A. Montes (Flickr: Boeing 787 Dreamliner N787BX) [CC BY 2.0 (http://creativecommons .org/licenses/by/2.0)], via Wikimedia Commons.

Figure 1.3: Playing Pool
Source: Nic McPhee [CC BY 2.0 (http://creativecommons.org/licenses/by/2.0)], via Wikimedia Commons.

Figure 1.4: Playing Pinball
Source: Gregg Tavares [CC BY 2.0 (http://creativecommons.org/licenses/by/2.0)], via Wikimedia Commons.

the difference and apply the rules of pool, even when they are playing a different game of pinball. Project managers and PMOs need to understand that the mechanical mindset of traditional approaches can be limiting and insufficient to deal with a rapidly changing, uncertain, and unpredictable DANCE-world. To play pinball, you need to constantly sense, respond, adapt, and adjust (SRAA). Over a period with practice, you can decipher the dynamics and become a pinball wizard. But organizations are far more complex and hard to decode the DANCE that impacts them. They require a nuanced and context-sensitive application of process and methods and humility to understand you cannot control everything; in fact, in pinball when you try to control, you go into tilt, and it's game over.

Traditional project management and PMOs that do not recognize the difference between the need for a different approach for the two games are set up to scale for efficiencies. They overemphasize planning, prevention, prediction, and process, which enables organizations to achieve consistent results in a mostly simple to complicated world. Pointing out the distinction between complicated and complex is important. Figure 1.5 describes the complexity continuum.

A bicycle is simple; an aircraft engine is complicated, not complex; you can rely on a manual to identify all the components and their behavior in predictable ways. Now, think of a stubborn horse, a team member, or any living thing that is hard to predict or control is complex.

Dealing with Complexity

Organizations, project environments, social networks, industries, governments, traffic flows, nature, and biological systems are examples of complex adaptive systems.

A complex adaptive system is a system in which a perfect understanding of the individual parts does not automatically convey a perfect understanding of the whole system's behavior. Complexity science is an interdisciplinary field that delves into understanding the properties of complex systems. Following are some of the properties of complex adaptive systems summarized from the work of De Toni and Comello, viewed through the lens of application to project management and PMOs:

Self-organization and emergence. Spontaneous emergence of new structures from the bottom-up; the behavior of the system differs from the simple addition of its parts; components self-organize to produce capabilities and outcomes that are neither obvious nor predictable.

The Impossibility of forecast. Boundary state between predictability and unpredictability, impossible to forecast precisely.

Simple   Complicated   Complex   Edge of   Chaos
Chaos

Figure 1.5: Complexity Continuum

Power of connections. Everything is connected to everything else; identifying and strengthening crucial connecting nodes is key.

Circular causality. The effect has a feedback on its own cause; cause and effect are nondeterministic.

Hologrammatic principle. The part is in the whole, and the whole is in the part.

Try and learn. Trial and error is the only way to learn and adapt.

Edge of chaos. State of dynamic balance between order and disorder.

The DNA of strategy execution has evolved based on these principles, and in this book, we will delve into the practical application of these properties to next-generation project management and PMOs.

Organizations and project environments are complex adaptive systems. It is necessary to create awareness and classify project complexity to determine the appropriate management approach. Projects move and shift between simple, complicated, complex, and chaos along the complexity continuum. The DANCE elements can be present on any project, regardless of size or cost or other elements and can either impact individually, or interplay in combination. What may seem like a minor element can have a ripple effect with unintended consequences on any project or program.

Key Question – How to Redesign and Organize for Increasing DANCE and Complexity?

Traditional approaches evolved to iterative and incremental with progressive elaboration and rolling wave planning, but today we are no longer dealing with waves—it is more like a tsunami. The volume, variety, and velocity of changes are intensifying the DANCE. We are in a constant state of flux, and the challenges are less predictable. Digitization accompanied by breakthrough technologies like artificial intelligence and machine learning is creating multiple inflection points. Though inflection points have always been a part of business reality, the frequency of inflection points has rapidly increased and is disrupting established businesses. Projects in today's environment are akin to building the plane while you are flying it. The path forward is not clear, and the unforeseen and unexpected are happening at a faster pace, causing a greater degree of turbulence. Earlier, you could predict the ebb and flow of order and chaos, or equilibrium and disequilibrium, but today's world is characteristic of unpredictable instability and disequilibrium.

Figure 1.6 lists the traditional approaches we have relied on for the simple to complicated. The question is, how do we thrive in an increasingly DANCE-world where the problem is unknown or unknowable? How do we deal with wicked problems that are difficult because of incomplete, contradictory, and changing requirements that are difficult to recognize, and the solution requires experimentation and learning?

This book delves into the question of what approaches are needed in the realm of complicated to complex, and how to balance between them.

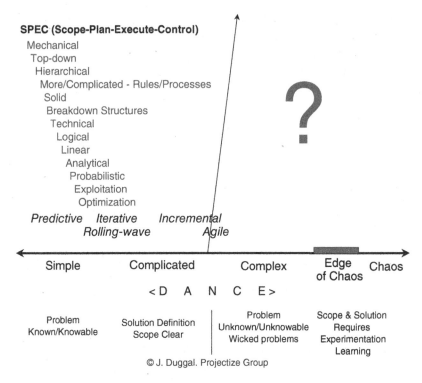

Figure 1.6: Dealing with the Simple to Complex

THE EVOLUTION OF MANAGEMENT FOCUS

The evolution of management focus can be traced in three paradigms, starting with efficiency—faster, better, and cheaper processes, products, and services for most of the nineteenth and twentieth centuries. In the latter part of the twentieth century in the 1980s and 1990s, the focus shifted toward effectiveness—extracting value and benefits, instead of efficiency for the sake of efficiency. Today's digital and hyper-connected world is based on experience—starting with the end user and customer experience, it builds on the optimization of efficiency and effectiveness. The next generation goes beyond with a focus on the transformative purposeful impact of the whole. Figure 1.7 highlights the key distinctions between efficiency, effectiveness, experience, and the next-generation evolution of project management and PMO.

Which category does your organization reflect? In our observation, many project management and PMO approaches are still stuck in efficiency, and some are moving toward effectiveness, while the awareness and practice of experience focus are rare. This book will delve into the above characteristics and discuss how all three are

	EFFICIENCY Traditional/ Foundational	EFFECTIVENESS 2nd Generation Project Mgt. & PMO 2.0	EXPERIENCE 3rd Generation Project Mgt. & PMO 3.0	IMPACT Next Generation?
Mindset	Mechanical (factory-oriented views organization as complicated machine; linear, siloed thinking)	Systems (connected, integrative thinking)	Organic (views organization as complex adaptive system; non-linear, adaptive and holistic thinking)	Intuitive (hybrid of organic/ mechanical-augmented with automation and intelligence)
Purpose & Focus	Execution and delivery	Results, benefits and value Strategic decision-support and prioritization	Strategy-execution – linking strategy and execution, with a stakeholder and customer focus – customer first, customer experience, customer success, and customer creation, and retention orientation	Strategic-execution – long-term gaze; shift to network benefits, from individual customer benefits; greater value and impact, while maintaining customer commitment and loyalty; sustain-ability and legacy
Role	Scope, Plan, Execute, Control (SPEC) Standardize (do it right and consistent)	Service, support, coaching, consulting Strategic decision-support Prioritize (do the right things)	Force multiplier - facilitate, expedite, connect and link; Enable agility and innovation;	Sense, Respond, Adapt, Adjust (SRAA)
Approach	Controlling and planning (Top-down) Predictive and analytical Risk intolerance Failure is not an option	Collaborative Iterative and incremental Agile/Lean methods Risk tolerance to a degree	Adaptive Experimentation (trial & error) Sensing & perceiving (new & different lenses) Customer engagement and collaboration Smart risk-taking	Integrative, intuitive and emergent (bottom-up - self-organization) Designing, architecting, and choreographing
Governance	Compliance orientation (monitoring and control – rigid processes; forced compliance)	Delivery orientation (support and collaboration – flexible processes; voluntary compliance)	Business and customer orienta-tion (responsive & adaptive processes; self-regulating and desire based governance)	Network orienta-tion designed to optimize the whole (intuitive, automated. and augmented governance)

Figure 1.7: Management Evolution—From Efficiency and Effectiveness to Experience and Impact
Source: © J. Duggal. Projectize Group.

	EFFICIENCY Traditional/ Foundational	EFFECTIVENESS 2nd Generation Project Mgt. & PMO 2.0	EXPERIENCE 3rd Generation Project Mgt. & PMO 3.0	IMPACT Next Generation?
Measurement & Success Criteria	Compliance and certification Deliverables and outputs On-time, on-budget delivery	Benefits and outcomes Customer satisfaction	Business value and impact Customer creation; retention; Net promoter score (NPS) Learning and innovation	Long-term impact; sustainability and legacy
Ownership & Accountability	Tasks, outputs, and deliverables	Delivery of benefits, and outcomes	End-to-end customer success and impact	Long-term impact; sustainability and legacy
Mantra	Faster, better, cheaper	Optimization of benefits and value	Agility and adaptive	Intuitive and intelligent

© J. Duggal. Projectize Group

Figure 1.7: (*Continued*)

important, but it is crucial to balance and evolve. Each generation builds on the previous, to evolve to the next generation.

AGILE DANCE AND BEYOND

Recognizing the limitations of traditional approaches and the inability to manage the DANCE effectively, the software development community developed agile. Agile resonates with many who are engaged in DANCE type of projects and understand they are playing a different game and need a different approach. Agile has become mainstream and is spreading outside of software development into other areas of management as well.

To apply agile effectively, you must rewire the organizational mindset and culture to understand and practice the agile manifesto—individuals and interactions over processes and tools; working software (products) over comprehensive documentation; customer collaboration over contract negotiation; responding to change over following a plan (agilemanifesto.org). And the agile principles—highest priority is to satisfy the customer; welcoming changing requirements; delivering working products (software) frequently; simplicity; self-organizing teams; reflection among others. While agile has been an antidote and has transformed many teams and organizations, there are many who continue to struggle. The challenge is as many have jumped on the agile bandwagon and they are doing agile; they are not necessarily prepared for it. They are excited about trying agile methods and techniques, but the way they practice them is with an old mindset. As a result, they get more of the same. As Stephen Denning and others have pointed out, there is a difference between "doing agile" and "being agile." Before you can do agile, you must be agile; otherwise, it is faux agile.

The application of agile is primarily focused on execution and process agility. What about the other aspects of strategic agility? Besides agile methods and processes, project management and PMOs have to address multidimensional and contextual agility in other areas of linking strategy, governance, measurement, connection and communications, organizational change management, learning, and innovation.

This book aims to link execution to strategy and focus on both execution agility and strategic agility. The ideas on these pages will help you with agile transformation and to integrate agile in a holistic way. This book goes beyond, and looks at the nature of complex adaptive systems and complexity to prepare your organization for true agility. It will challenge you to anticipate what's next after the predictive, iterative, incremental, and agile approaches to be better prepared for the persistent DANCE and turbulence.

APPROACH

Question Everything

In the world of efficiency and effectiveness, we could survive and prosper by finding many alternatives to solve problems. Techniques like brainstorming, which aim to come up with as many solutions as possible, are still popular. You can quickly brainstorm multiple solutions; the challenge is that you could be working on the wrong question. The problem is not the answers but the questions. How do we know if we are asking the right question? How do we learn to ask better questions? To get better at strategy execution, you must challenge yourself and your teams to ask beautiful questions. Warren Berger, in his book, *A More Beautiful Question*, explains, "A beautiful question is an ambitious yet actionable question that can begin to shift the way we perceive or think about something—and that might serve as a catalyst to bring about change." Each chapter starts with leading questions. These questions will, of course, be addressed, but I hope that by the end of the book you will have more beautiful questions that provide new insights and open new doors. Like Larry Page, CEO of Alphabet, we need to learn to ask 10× questions that require answers that have 10 times the impact of previous solutions. You will learn to frame your challenges with better questions and continue to challenge the status quo and seek new perspectives with the power of questions.

> "Problems that remain persistently insoluble should always be suspected as questions asked in the wrong way."
>
> *Alan Watts*

Focus on the Purpose of Business

How would you respond to the question, "What is the purpose of business?" Most people respond, "to make money … increase shareholder value … to provide customer value … to make a sustainable impact … etc."

Way back in 1954, Peter Drucker responded:

> "There is only one valid definition of business purpose: **to create a customer**. The customer is a foundation of a business and keeps it in existence. The customer alone gives employment. And it is to supply the customer that society entrusts wealth-producing resources to the business enterprise."

He wrote these words in his book *The Practice of Management,* as far back as 1954, it is only now we are seeing that organizations are getting this. Think about it: if you don't have a customer, you don't have a business. That's why we use the services of companies like Google, Facebook, and many others for free; their job is only to create new customers. Once you have loyal customers, there are many ways to monetize. Drucker's words are echoing as the Copernican revolution in management is shaping up. As observed by author Stephen Denning, in the nineteenth and twentieth centuries, the firm was in the center, and the customer was on the periphery, just as before Copernicus in 1543 it was believed that the Earth was the center of the universe. With increasing choices and open access to products and information, it is hypercompetitive, and businesses can't succeed without putting the customer at the center, to make their products and services go viral, and to create and keep customers.

We hear strategy execution leaders like Jeff Bezos of Amazon say, "We see our customers as invited guests to a party, and we are the hosts. It's our job every day to make every important aspect of the customer experience a little bit better." Shifting to a customer-centric perspective can have a profound impact, as one of the nonprofit healthcare organizations we worked with started referring to patients as customers. This simple change precipitated a mind shift, and instead of viewing people as a set of symptoms, it began treating them differently.

The approach in this book is to start with the customer at the center of everything you do from a strategy execution standpoint in project management and PMO. Start with the question, "Who are our real customers, end-users, and stakeholders?" and develop deep customer first, customer experience, customer success (helping customers easily adopt and succeed with our products, services, and systems), and customer creation and retention obsession. The questions we will be looking at are, how do you better organize and manage projects in this customer-centric, customer-first, customer-success world? Is it possible to create a project management and PMO platform that connects, facilitates, and enables an end-to-end customer experience?

Apps versus Operating System

We are constantly looking for the new cool app, the next shiny tool, template, or best practice that is perhaps the silver bullet that is going to solve our problem. The

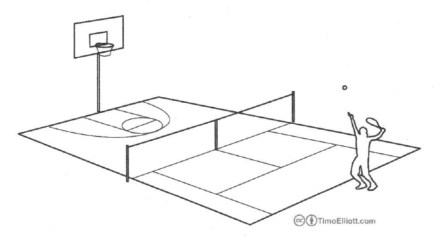

Figure 1.8: Which game are you playing?
Source: © TimoElliott.com.

challenge is that we can acquire the app, but it might not work because the operating system (OS) is incompatible. You cannot run an iOS app on Android or vice versa unless it is designed for both. It is important to understand this distinction between apps and OSs. Before you can acquire or use the app, you need to prepare the organization with the right OS. Unfortunately, we see a rush for the next app, method, tool, or technique, without consideration for compatibility with the organizational OS—the prevailing mindset and related context and culture that is needed for it to flourish. It is like rolling out the latest tennis gear, without realizing that your organization is designed for basketball—it is only going to add to the confusion as illustrated in Figure 1.8. Next-generation leaders focus first on the operating system of the organization, preparing and rewiring the mindset of the players, before implementing the methodology, tools, templates, or best practices. This book aims to do that by focusing more on the OS level, besides providing the apps.

Process versus Behavior

Based on a industrial factory mindset, most organizational efforts are geared toward processes and systems and not the people who use them. Our ongoing project management and PMO survey since 2005 highlights the fact that PMOs spend 85 percent of their time on developing, documenting, standardizing, refining, training, and measuring process and methods. Less than 10 percent proactively plan for the impact of organizational and behavioral changes that these processes cause. As some have quipped, that process is fine until people get involved because it is the people who are involved in designing, executing, misusing, or subverting the processes. Wherever it is possible, strive to remove the human from the process with automation of repeatable processes. For the rest of the processes, our approach in this book is to focus on the human, and the interaction between process and behavior. For each process, we will consider the behavioral aspects and emphasize the balance between the two. While process and methods are important, it is the people who have to adopt the processes, and adapt to benefit from them.

Simplicity

"Our enemy is not the competition; it is the unnecessary complexity in our processes," remarked an executive. This observation is a reality that resonates with many, as in the project management and PMO world we are guilty of making things more complicated then they need to be. As Einstein noted, "It is easy to make things complicated; it is hard to make them simple." We will strive hard to simplify and exploit unrecognized simplicities. As Andy Benoit observed, "Most geniuses—especially those who lead others—prosper not by deconstructing intricate complexities but by exploiting unrecognized simplicities." This is the heart of our approach, and simplicity is the foundational theme that will echo throughout the ideas, models, methods, and tools discussed in this book. In Chapter 11 we take an in-depth look at this topic.

Holistic

In an ideal world, the organizational mirror would be whole, in which you could see a coherent corporate strategy and structure, mission, vision, values, strategies, and objectives. The strategy is linked to execution and operations. There is an aligned and prioritized portfolio, with well-defined programs and projects, and everything is linked and transparent. But the reality is messy, and typical organizations are fractured and not as coherent and organized—even more so in today's DANCE and disruptive world.

To execute strategy effectively, you have to think holistically. As Peter Senge, author of the seminal book, *The Fifth Discipline,* in his work on systems thinking wrote, "Business and human endeavor are systems...we tend to focus on snapshots of isolated parts of the system. And then wonder why our deepest problems never get solved." The role of project management, PMOs, and strategy execution leaders should be to connect and build bridges, find commonalities, reduce redundancies, and spread ideas that work.

The project world is characteristic of reductionism and breaking things down, and the heart of the project management approach is breakdown structures. While it makes sense to break things down to better plan and manage, the challenge is bringing them back together. Even when you try to integrate, due to the cracks caused by the breakdown, it is hard to patch and get a holistic perspective to optimize the whole.

Identifying the foundational DNA elements across the management of business and related projects, programs, portfolios, and PMOs provides insights and opportunities to connect and optimize the whole. It provides the opportunity to leverage the complex systems property: *the part is in the whole, and the whole is in the part.*

In this book, we will take a holistic perspective; instead of separating things, we will try to bring them back together. While there is value in separating portfolios, programs, projects, PMOs, and organizational project management (OPM), we will look at them holistically by looking at the DNA level and identifying the common characteristics and themes that apply across the board. In the spirit of the 'whole' we will use project and program, and project management (in a broad organizational context) and PMO interchangeably; many of the ideas overlap and apply when you look at things at the foundational DNA level.

BACKGROUND

The ideas and insights discussed in this book are based on over 17 years of experience in designing and implementing organizational project management and PMOs in different organizations around the world. In our practice at the Projectize Group, we have worked with a few thousand people from hundreds of organizations since 2000, in leading Next-Generation PMO and Portfolio Management, Managing the DANCE, and Leadership seminars for the Project Management Institute (PMI) and various organizations and governments around the world. As facilitators, consultants, participant observers, and curious practitioners in the pursuit of next-generation approaches to strategy execution, we have not only observed, researched, discussed, and debated these ideas from multiple perspectives but, more importantly, implemented them. As best-selling author and scholar, Nassim Taleb said, "Instead of putting theories into practice, create theories from practice." We have applied, learned, and evolved these practices in many organizations in different industry verticals.

The stories and examples highlighted in the book are from established Fortune companies to Silicon Valley start-ups, and government organizations and nonprofits around the world. Also, we have continued to conduct and compile survey-based research since 2005 that continues to provide rich data from a highly selective sample of over 1600 managers, executives, and PMO leaders. Ongoing interviews with executives and PMO leaders have provided deeper insights and validation of the ideas.

This book will provide new perspectives and related tools for project, program and portfolio managers, PMO leaders, product owners, and executives, and anybody interested in the next generation of strategy execution.

How Can You Use This Book?

The impact of digitization and disruption and the intensification of the DANCE requires a rethink and reset of how we organize and manage. This book will present a holistic approach by decoding the DNA of strategy execution and provide new perspectives and practical applications, tools, and examples for each of the DNA elements. Projects and Programs are the vehicles for strategy-execution and this book can be used to:

- Develop a playbook for effective strategy execution.
- Improve your effectiveness as next-generation managers, leaders, owners, and executives.

- Gain new perspectives and insights to develop skills necessary to thrive in today's turbulent world.
- Build a next-generation project management and PMO platform for strategy execution.
- Decode the core elements of the DNA of management and strategy execution.
- Prepare your organization to lead and implement agile transformation in an effective way.
- Redesign and transform project management and PMO, and apply next-generation ideas to take your PM capabilities and PMO to the next level.
- Balance the need for foundational management discipline with the need for agility, creativity and innovation, and achieve the rigor but without rigidity.
- Evolve to an organizational project management (OPM) and PMO Center of Excellence (CoE).
- Foster effective measurement, feedback and learning for continuous improvement and innovation.
- Find ways to connect and bridge the gaps across organizational silos.
- Challenge conventional and unproductive management practices.
- Change negative project management and PMO perception.
- Measure and increase business value and impact, and gain buy-in, with raving customers and promoters for project management and PMO.
- Improve organizational project management (OPM) and PMO maturity by developing intelligence in each of the DNA areas of strategy execution.
- Improve overall organizational effectiveness and innovation capabilities.

The next chapter will start by providing the context as to why and how to reframe the organizational operating system and mindset for agility, why project management and PMO efforts fail, and what are the next generation ideas to evolve and adapt. In Chapter 3 we will decode the DNA of strategy execution. In the following chapters, we will look at each element of the DNA in detail, starting with strategy in Chapter 4, and execution in Chapter 5. The other elements—governance, connect, measure, change, and learn—will be covered in Chapters 6 through 10. Chapters 11 and 12 focus on enabling aspects to apply the DNA of strategy execution with simplicity and balance to thrive at the edge of chaos.

2

AGILITY: RIGOR WITHOUT RIGIDITY

"Success today requires the agility and drive to constantly, rethink, reinvigorate, react, and reinvent."

Bill Gates

Leading Questions

- How do we reframe the organizational operating system and mindset?
- How do we get project management and PMO ready for DANCE and disruption?
- Why do half of PMOs fail, and what can we learn from them?
- How do we distinguish traditional or foundational versus next-generation approaches?
- How do we know which approach is right for us?
- Is it possible to be agile by focusing on both speed and stability?
- How do we find the sweet spot between the need for rigor but without the rigidity?
- What are the seven keys to successful project management and PMOs?

A 100-year-old iconic company admired for decades as the most innovative company in diverse industries from automotive to electronics, healthcare, and consumer products appoints a new CEO in 2000. His focus is on execution and operational efficiencies as he implements Six Sigma to streamline work processes and focus on cost efficiency and quality control. The company was 3M, well-known for innovations like Scotch tape and Post-it notes. What do you think would be the impact of process rigidity on an innovative company like 3M?

While there were efficiency gains and operating margins grew from 17 percent to 23 percent, there was a dramatic fall in the number of innovative products developed in those years. According to *Fast Company* magazine, 3M fell from number 1 in 2004 to number 7 on the most innovative companies list. James McNerney, the CEO of 3M from 2000 to 2005 applied Six Sigma across the board. Six Sigma is process heavy and aims to remove variability to avoid errors and increase predictability. Developed at Motorola, Six Sigma was adapted by Fortune companies like GE, Honeywell, and many others in the 1990s. Processes like Six Sigma are about consistency and control and work well in a mechanical world, whereas the DANCE and disruption are characteristic of difference, failure, disorder, learning, and mutation.

Project management and PMOs face a similar challenge as they are perceived as top-down, controlling, and bureaucratic, stifling innovation in an increasingly DANCE and disruptive world. On the other end of the extreme, there are start-ups and companies with very little or no process, and chaotic, free-for-all cultures with no boundaries and constantly shifting focus. Any management process, whether it is Six Sigma or project management or project management office (PMO), can be beneficial, but the challenge is knowing how much is right for your organization and business. Failure stems from either too much or too little governance and process. The challenge is that to thrive, you need to find your sweet spot between the two extremes.

In his book, *Seeing What Others Don't: The Remarkable Ways We Gain Insight,* cognitive psychologist Gary Klein argues:

> Organizations are preventing insights by imposing too many controls and procedures in order to reduce or eliminate errors. Organizations value predictability and abhor mistakes. That's why they impose management controls that stifle insights. If organizations truly want to foster innovation and increase discoveries, their best strategy is to cut back on the practices that interfere with insights, but that will be very difficult for them to do. Organizations are afraid of the unpredictable and disruptive properties of insights and are afraid to loosen their grip on the control strategies.

In this chapter, we address the question of how to achieve true agility by finding your sweet spot between the two extremes. It is important to prepare the right operating system before we delve into decoding and developing the DNA of strategy execution for project management and PMO practices.

HOW TO REFRAME THE ORGANIZATIONAL OPERATING SYSTEM AND MINDSET

> "The most common source of management mistakes is not the failure to find the right answers. It is the failure to ask the right questions.... Nothing is more dangerous in business than the right answer to the wrong question."
>
> *Peter Drucker*

The next generation depends on questions. In the past, we relied on answers, but in today's world as the DANCE intensifies and uncertainty increases, the value of questions goes up, and the value of answers goes down. It is harder to understand the problem than to have answers that are no longer relevant. How can you expect different outcomes with the same questions? The challenge is to learn to ask better questions. While solutions tend to converge and pacify, questions have the opposite effect. They agitate and cultivate a mindset for agility, constantly challenging the status-quo, seeking diverging perspectives, and pushing to the edge. To better prepare for the DANCE and disruption, you have to reframe the organizational mindset and create a culture of curiosity. Questions are like lenses that provide a multiplicity of perspectives and unlock new doors.

The PMO can take a leading role in fostering a question-based culture across the organization by holding up the mirror and reflecting the right questions in a multidisciplinary way, bridging silos and reconnecting different aspects of the DNA of strategy execution. Traditional project management and PMOs are typically focused on finding answers and providing solutions for execution. The next-generation PMO should start by challenging if you are working on the right problems and projects by using different questioning techniques.

In our next-generation PMO seminars, we start with a question-storming exercise. It is a divergent process designed to come up with a multiplicity of perspectives with the power of questions, as opposed to brainstorming, which is convergent honing on multiple solutions to one problem. Hal Gregersen, executive director of the MIT Leadership Center and senior lecturer at the MIT Sloan School, explains:

> Regular brainstorming for ideas often hits a wall because we only have so many ideas. And the reason we hit that wall is we're asking the wrong questions.... It's perfect to step back and say: Okay, question storming time. And what do I mean by that? If I'm with the team: Grab a flip chart. Have someone be the scribe and have the team generate at least 50 questions about the problem that we're stuck on. Number those questions. Write them down up there, when they are being written down other team members are paying attention and thinking of a better question. Do it again and again and again. But what we discover is that when people then step back and do that kind of question storming, list a long series of questions and they do it collectively where they can see the questions and generate new ones, it actually gets them closer to the right question that will give them the right answer. And that's where question storming compliments traditional brainstorming.

Hal has developed a methodology—catalytic questioning, an alternative to traditional brainstorming—business leaders can use to drive change in their lives, workplaces, and communities. It takes just five simple steps:

1. Gather employees around a writing surface.
2. Choose the right problem.
3. Engage in pure question talk.
4. Identify the "catalytic" questions (questions that hold the most potential for disrupting the status quo). Focus on a few questions that your team honestly can't answer but is ready and willing to investigate. Winnow your questions down to three or four that truly matter.
5. Find a solution.

Question storming is also referred to as *Q-storming* by Marilee Adams, in her book *Change Your Questions, Change Your Life*. She elaborates that the goal in Q-storming is to generate as many questions as possible. The expectation is that some of these questions will provide desired new openings or directions. Typically, questions open thinking, while answers often close thinking. Q-storming is based on three premises: (1) great results begin with great questions; (2) most any problem can be solved with enough of the right questions; and (3) the questions we ask ourselves often provide the most fruitful openings for new thinking and possibilities.

In the Q-storming exercise, groups of stakeholders start with what questions we need to ask for the PMO to provide greater value to the organization, or what questions we should be asking to take the PMO to the next level.

Table 2.1 is an unedited Q-storm from one of our PMO facilitation workshops.

After a Q-storming exercise, we typically hear comments like, "This was an eye-opener ... we didn't know we had to think about all these things ... before I came to this session, my focus was so narrow ... now we know we are not even thinking about the right things. ..."

Project managers and PMOs must shift from a fixation on the solution toward learning to ask better questions that help reframe the problem and provide a different point of view and open new doors. The questioning approach will raise the quality of planning, risk management, and surfacing assumptions, among other things. Questioning is an art and skill that has to be learned and developed. Table 2.2 shows sample questioning techniques compiled from different sources.

Throughout the book, we will be using questions to open new doors and keep seeking new insights to develop intelligence in each of the elements of the DNA.

> "I would rather have questions that can't be answered than answers which can't be questioned."
>
> *Richard Feynman*

One of the questions to start with is why half of PMOs are not perceived as successful? According to the Project Management Institute's Pulse of the Profession, 2017, 71 percent of organizations have a PMO, up from 61 percent in 2007. Even though PMOs have become a common organizational fixture in many organizations, the success rate has not gone up. According to Projectize Group's survey-based PMO research (2005–2017), 52 percent of PMOs are not perceived as successful by key PMO stakeholders. 39 percent responded that the relevance or existence of their PMOs has been seriously questioned. These findings are echoed by other reports from various organizations over the years, including a multiyear PMO study by Brian Hobbs and Monique Aubrey, *PMO: A Quest for Understanding*.

Often, you learn more from questioning why something doesn't work than from why it does. One of the exercises we conduct in our practice is to ask the key stakeholders an important question: *What can you do in your PMO to ensure it does not succeed?* This is backward thinking, a technique, explained by Peter Bevlin, in his book *Seeking Wisdom*. Instead of asking how we can achieve a goal, we ask the opposite question: What don't I want to achieve (non-goal)? What causes the non-goal? How can I avoid that? What do I now want to achieve? How can I do that?

Table 2.1: Sample Q-Storm of what we need to do to take the PMO to the next level

What is the need for the PMO?	What's our appetite for change?	What is the right org structure for the PMO?
What if we did not have a PMO?	How will we get feedback from our customers?	What support do we have and need?
What should PMO stop doing?	How should we train our resources?	Who are the most influential and key players?
Why do people not like us?	What PMO model is right for us?	How do we say "no" and make it stick?
What pain are we causing?	How will we ensure we're aligned with overall biz strategy?	What will make project managers happy?
How do we define PMO success?	What is the overhead on our projects?	What behavior do we want to encourage?
How will we measure success?	How do we put structure in place but remain flexible?	What authority does the PMO have?
Are we measuring the right things?	How do we evolve as the organization changes?	Are we clear on our business strategy?
What processes should be improved?	What do our customers value?	How do we engage and get people excited about project management?
What tools are right for us?	How do we market ourselves?	How do we identify and reduce disconnects?
How do we keep the PMO relevant?	How can we gain the support we need from different stakeholders?	How can we make ourselves invaluable?
How do we measure PMO value and return on investment (ROI)?	If we started from scratch, how would we go about it?	What relationships do we need to build?
Who are our customers?	How can we drive the right behaviors?	Are we listening?
What are we doing well?	What is our impact on people?	Do we know what our capacity is?
What are we not doing well?	How are we changing how people feel about performing their jobs?	Are we moving the needle?
What are our organization's pain points?	What does an ideal PMO look like?	Are we stuck in status quo?
What can we adapt from other successful PMOs?	What is the pain that the PMO should address?	What should be the roadmap for the PMO?
What does failure smell like?	Why do projects fail?	What roadblocks and dependencies do we see coming?
Do we have the right resources?	What do customers want?	How do we develop and retain talent?
What environmental factors influence us?	How do we prioritize what we need to do?	How can we connect and communicate better?

Table 2.2: Sample Questioning Techniques

Technique	Description	When to Use
Five Whys [Toyota]	Investigative method of asking the question "Why?" five times to understand what has happened (the root cause). Each question forms the basis of the next question. The technique was formally developed by Sakichi Toyoda and was used within the Toyota Motor Corporation during the evolution of its manufacturing methodologies. The tool has seen widespread use beyond Toyota and is now used within Kaizen, lean manufacturing, and Six Sigma.	To explore the cause-and-effect relationships underlying a particular problem. To shift your perspective and point of view—helps in re-framing problems.
Why–What If–How [Warren Berger]	A framework designed to help guide one through various stages of inquiry—because ambitious, catalytic questioning tends to follow a logical progression, one that often starts with stepping back and seeing things differently and ends with taking action on a particular question.	A model for forming and tackling big, beautiful questions that can lead to tangible results and change.
How Might We [Min Basadur/Sydney Parnes]	The specific form of questioning with the words "how might we" designed to spark creative thinking and freewheeling collaboration. By substituting the word might instead of can or should, you are able to defer judgment, which helps people to create options more freely and opens up more possibilities. This technique is also extensively used at IDEO. Tim Brown, CEO of IDEO, explains, "The how part assumes there are solutions out there—it provides creative confidence. Might says, we can put ideas out there that might work or might not—either way; it's okay. And the we part says we're going to do it together and build on each other's ideas."	For creative problem solving and generating multiple options and possibilities.

Table 2.2: (Continued)		
Technique	**Description**	**When to Use**
Propelling Questions and Can-If [Adam Morgan and Mark Barden]	A propelling question is one that has both a bold ambition and a significant constraint linked together. It is called a propelling question because the presence of those two different elements together in the same question does not allow it to be answered in the way we have answered previous questions; it propels us off the path on which we have become dependent. For example: How do we win the race with a car that is no faster than anyone else's? (Audi) How do we build a well-designed, durable table for five euros? (IKEA) Shift from We Can't Because … to We Can-if … to instill a sense of how something could be possible, rather than whether it would be possible.	To achieve a bold ambition, but with a significant constraint that helps to break path dependence, and spur entirely new kinds of solutions. For constraint-driven innovation.
Then What [Warren Buffett]	To elicit consequences—whenever someone makes an assertion to you, ask, "And then what?" According to Warren Buffett, "Actually, it's not such a bad idea to ask it about everything. But you should always ask, "And then what?"	To help drive focus on consequences.
What Must Be True [Roger Martin]	To specify what would have to be true for the option or choice to be right. Can be used as a collaborative approach to surface the assumptions and understand the convincing logic of the options.	To surface an assumption and turn it into an advantage.
If-Then	To highlight any conditions that need to exist to achieve outcomes—dependencies, interfaces, policy considerations, resources, market factors, and other important conditions to make if-then logic valid.	To communicate project plans, progress reports, and expected results or outcomes. To highlight critical assumptions and risk factors.

What can you do in your PMO to ensure it does not succeed? (non-goal)

Typical responses we see listed are ...

Don't communicate the purpose of the PMO
Disconnect between strategy and execution
Lack of business understanding and focus
Require too many mandated bureaucratic processes
Don't explain why they have to follow PMO processes
No training or support
Don't define roles/responsibility clearly
Inflexibility and one-size-fits-all policies
Not addressing the current needs and pain points of the organization
Set high expectations and unachievable goals
Punish people for not following processes
Reward bad behaviors
Don't pay attention to organizational culture or politics
Don't plan for organizational change management
Lack of stakeholder and customer engagement
Don't care or plan for stakeholder buy-in for the PMO
Stumbling block for new ideas and innovation
Slow down agility
Lack of understanding of reality of DANCE and disruption

An interesting outcome of this exercise is that, invariably, it surfaces the following top reasons why organizational project management or PMOs fail:

- Unclear purpose
- No buy-in
- Perception of more red tape, bureaucracy, and overhead
- Quick-fix to deep-rooted problems
- Project management policing
- Too academic and far from reality—professionalism and quality for its own sake
- Veneer of participation and hidden agendas
- Politics and power struggles
- High expectations and fuzzy focus
- Hard to prove value

The preceding reasons are prevalent because traditional approaches are skewed toward a top-down, controlling, policing, rigid, and bureaucratic project management and PMO based on a mechanical mindset. They do not recognize the DANCE and are unaware that they are playing a different game, which requires a different approach as discussed in the first chapter. It is important to understand fundamental distinctions between traditional or foundational approaches, and evolving start-up, organic, and agile approaches.

DISTINGUISHING TRADITIONAL VERSUS EVOLVING APPROACHES

Figure 2.1 highlights the distinctions between foundational/traditional versus evolving/agile/adaptive organizational and PMO approaches. You can check either one of the characteristics and add the number of your checks at the bottom to get a score that will highlight whether you lean more toward traditional/foundational or evolving/adaptive.

Foundational/Traditional	Evolving/Adaptive
Mechanical or Factory mindset (machine-oriented)	Organic mindset (knowledge-oriented – non-linear & connected)
Focus mostly on execution and delivery	Focus on strategic decision support, business value, end-user & customer experience
Focus on science of management (technical expertise)	Focus on art and craft of management (organizational savvy)
Emphasis on monitoring and control	Emphasis on support and collaboration
Provides tools similar to a precise "map" to follow	Provides tools similar to a "compass" that show the direction
Rigid and formal structure, process, and governance standards, methods and processes (one-size fits all)	Responsive, flexible and flat structure, agile governance methods and processes (context dependent)
Extrinsic approach: Top-down, chartered, selected, outlined	Intrinsic approach: Evoke, provoke, encourage creativity, experimentation & learning
Process driven: Focus on process & methodology compliance	Customer & business driven: Focus on value, experience, & impact
Emphasis on rules; follow rules	Based on guiding principles; Follow rules and improvise if needed
Seek stability & global consistency	Seek agility and innovation
Focus on the "WHAT" (task management)	Focus on "Why" & "WHO" (stakeholder/relationship management)
Focus on efficiency: Manage inputs & outputs	Focus on effectiveness & experience: Ownership of results & outcomes
Exploitation	Exploration
Measure & track compliance; certification & delivery	Measure & track benefits, value, experience, & impact
Foundational/Traditional Score = ?	**Evolving/Adaptive Score = ?**

Figure 2.1: Traditional versus Evolving Approaches
Source: © J. Duggal. Projectize Group.

Figure 2.1 highlights the fundamentally different approaches to setting-up and managing organizations, based on a mechanistic view of foundational or traditional aspects of standards, stability, and rigidity versus flexibility, agility, and responsiveness in the evolving, organic mindset. On the one hand, there is a need to establish rigor with a sound governance structure of methods and processes; on the other hand, there is a demand for freedom and flexibility. This is indeed a primal paradox between the need for discipline and freedom at the same time. This dilemma hounds the successful implementation of project management and PMO processes. It surfaces the underlying friction and challenges long-questioned beliefs about control and the role of management. To bring about the discipline, you need the rigor of standards and processes. However, the rigor can turn into rigidity that restricts judgment and stifles creativity and tend to slow and restrain organizational agility and innovation.

The distinction also highlights false assumptions that are inherent that the right side is better than the traditional approaches of the left side, or vice versa. For example, start-ups can move and pivot with speed and agility, or established companies have consistency and predictability. After a certain point, start-ups struggle to maintain momentum, lose focus and become chaotic, while traditional companies become bureaucratic and rigid that hamper their ability to move. Traditionally, the purpose of PMOs is to implement standards and processes and are more skewed toward the left.

Another assumption is that executives have to make a trade-off and choose between one or the other, speed and agility, or standards and stability. The classic mistake is to treat it as an "either/or" choice, instead of "and/both" thinking that can identify links between opposing forces and generate new ideas. To practice this, you have to understand integrative thinking and apply the power of paradoxical frames. Roger Martin, in his book *The Opposable Mind,* defines integrative thinking as "the ability to face constructively the tension of opposing ideas and, instead of choosing one at the expense of the other, generate a creative resolution of the tension in the form of a new idea that contains elements of the opposing ideas but is superior to each." By acknowledging and combining the opposing elements, you temper the undesirable side effects of each element and enable new insights that integrate both elements.

Stanford professor Charles O'Reilly explains organizational ambidexterity as,

> the ability of an organization to compete in mature markets and technologies—where key success factors involve efficiency, incremental improvement and short timeframes; and simultaneously, to compete in emerging markets and technologies—where key success factors require flexibility, initiative, risk-taking and experimentation. Research indicates that the ability to do both of these at once is associated with long-term success.

WHAT IS NEXT-GENERATION PROJECT MANAGEMENT AND PMO?

Next-generation project management and PMO is ambidextrous; it recognizes the importance of both the elements and does not focus on either/or. It acknowledges the tension between the paradoxical elements but understands that marrying the two reduces the undesirable effects of each and leads to new insights and blended solutions. For example, make sure everything is planned for the go-live, but also remain flexible so that we can deal with last-minute requests from customers. This is a classic paradox that stumbles traditional PMOs, who do not recognize how planning and flexibility can positively reinforce each other. Continuous planning can help you be better prepared for customer changes and enable greater flexibility.

The next-generation PM and PMO is better prepared to deal with the DANCE and disruption because it pushes strategy execution to the edge, by constantly questioning, seeking, and learning. It learns how to juxtapose the opposing forces of the need for rigor and flexibility and use them to create a dynamic strategy execution environment. Next-generation PM has to use a bimodal approach or barbell strategy. Nassim Taleb, in his book, *Antifragile: Things that Gain from Disorder,* describes the barbell strategy by explaining that the barbell is meant to illustrate the idea of a combination of extremes to describe a dual attitude of playing it safe in some areas and taking a lot of small risks in others. Extreme risk aversion on one side, and extreme risk loving on the other, thus reducing the downside risk of ruin, while capitalizing on the positive risks. You can't plan for the unpredictable, but you can build shock absorbers, and increase absorptive capacity, by embracing and capitalizing on the paradoxes.

Which Quadrant Is Your Organization in Today?

In Figure 2.2, the rigor versus responsiveness matrix lists the characteristics of each quadrant from the two extremes of bureaucratic and start-up environments to the balanced next generation. Check each box that relates to your organization. Add the number of checks in each quadrant to determine where your company is today.

Rigidity can be defined by a heavy emphasis on formal structures and control, standard methods and processes, top-down governance with dictated authority, responsibility, and decision making. You are expected to follow precisely defined rules and procedures rather than to use personal judgment. In contrast, responsiveness means the project management practices or the PMO is responsive to the stakeholder and business needs. It emphasizes flexibility, adaptable and customized processes, and self-regulating governance with shared authority, responsibility, and decision making. A responsive PMO is tuned to the shifting business environment and can respond to changing priorities. For example, you may have worked hard to come up with a consistent project selection criteria model, but a responsive PMO would be open to adjusting and fine-tuning the model rather than stubbornly proposing a one-size-fits-all approach.

Typically, the idea of project management methodologies and PMOs conjures up images of bureaucracy and loads of unnecessary paperwork. In the Projectize Group survey referenced earlier, 72 percent of PMO stakeholders perceived their PMOs to be bureaucratic. Indeed, project management methodologies and PMO practices are often guilty of inflicting too much process, like requiring your project managers to complete two weeks of project documentation on a one-week project. It is akin

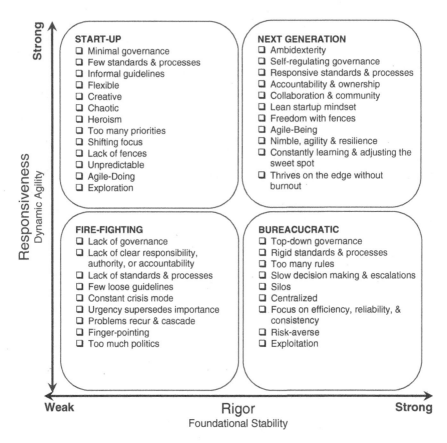

Figure 2.2: Rigor vs. Responsiveness
Source: © J. Duggal. Projectize Group.

to installing an elaborate security system on a cookie jar with a detailed process for removing cookies to prevent your child from eating too much sugar.

The traditional notion of control and mandated compliance can be an illusion. You may think you are gaining control, but what you are getting is paperwork and bureaucracy, while project managers and team members seek ways to undermine the processes. They may fill out the forms and templates to the letter to get the PMO off their backs, but not necessarily with the right spirit. Whether you are trying to bring up your child in a disciplined environment or you are trying to implement the discipline of project management practices, you go through a similar struggle. To bring about the discipline, you need the rigor of standards and processes. However, the rigor can turn into rigidity that restricts agility.

How to Achieve Balance

3M did manage to regain its groove. In July 2005 a new CEO, George Buckley, was appointed. He worked to preserve the benefits of Six Sigma's cost-cutting and efficiency-improvement efforts while simultaneously re-stimulating the creative and innovative juices at 3M. The way he managed to find the sweet spot was to exempt a lot of the research process from the more formal Six Sigma forms and reports. He found a way to balance by having process rigor where it made sense and toning it down in other areas. According to Buckley, "Invention is by its very nature a disorderly process.... You can't put a Six Sigma process into that area and say, well, I'm getting behind on invention, so I'm going to schedule myself for three good ideas on Wednesday and two on Friday. That's not how creativity works." By 2010, 3M had restored its innovative edge, and since 2010 it has continuously made it to the top 10 in Strategy & PwC's top innovator's list.

The need for rigor and freedom are opposing forces that cause friction between the PMO and its stakeholders. You have to seek the balance between the extremes of rigidity and responsiveness to varying degrees, as illustrated in Figure 2.3.

The very purpose of implementing project management and PMOs is to achieve discipline. Either extreme, rigid processes or flexibility with no standards, is not desirable. It is imperative to optimize the balance between two forces inherently at cross purpose.

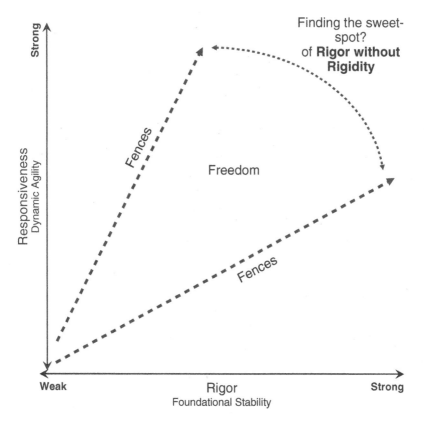

Figure 2.3: How to Achieve Balance
Source: © J. Duggal. Projectize Group.

As Peter Drucker said, "Most of what we call management consists of making it difficult for people to get their work done." The opportunity is to find the balance that makes the PMO responsive to the needs of the stakeholders, so it is easier for them to manage projects. For example, the PMO has to target the right mix of rules versus guidelines, instead of strictly defined "Ten Commandments," the PMO may require "Four Commandments and Six Suggestions."

Finding Your Sweet Spot

Use the rigor without rigidity figure to create a matrix and identify which quadrant your PMO is in. The next step is to identify the structures, standards, processes, and rules that are the backbone and required and have no leeway for flexibility. Each of the remaining processes should be adjusted toward responsiveness to create more dynamic processes that can adapt to changes and opportunities. Think about agile values of individuals and interactions over process, working software and products over documentation, customer collaboration over contract negotiations, and responding to change over following a plan, as you try to achieve balance. The appropriate location on the matrix will depend on a number of factors, including:

Business. The nature and criticality of business dictate the level of rigidity required. For example, a hospital or a nuclear power plant might have a more urgent need to establish standard policies and procedures than a consulting company.

Content. The content and relevant importance of processes, policies, and procedures determine which side of the grid they may lean toward. For example, security and privacy processes or project budget governance should be more rigid than configuration management.

Scope. The extent of rigor should be determined by the scope of the project. It is a good idea to define project classification criteria that helps to classify projects into simple, medium, and complex categories. For example, a simple project may need very limited process steps versus a complex project that may require more elaborate methodology steps.

Culture. More methodology implementations flounder from failing to adequately account for an organization's cultural factors than for any other reason. A culture's level of receptivity to change can make or break any effort to make improvements or achieve compliance. For example, it is easier to implement a project management methodology in a control-oriented culture versus a cultivation-oriented innovation culture where people abhor standards and processes.

Maturity. The receptiveness to processes depends on the overall organizational maturity of project management practices. Maturity will be based on the availability, quality, training, and support of the methods and processes that are provided. The greater the maturity chances, the better understanding and greater appreciation of the real intent of the processes and the greater the effort toward balance and voluntary compliance will be.

It should also be noted that over time the sweet spot may shift to varying degrees on the grid. This is why it is important to regularly rethink and redesign governance and related processes to strike a balance and achieve rigor without rigidity.

A PMO team from a U.S. multinational company expressed how excited they were a year ago when management had finally approved an enterprise PMO to work with the local and regional PMOs to establish global standards and procedures. The global PMO team worked hard in rolling out a project management methodology and standard templates with detailed procedures and mandated requirements. However, after a few months of completing their rollout, they were frustrated and could not understand why they were getting resistance and limited compliance.

Another PMO team of a global financial services company deployed a worldwide project management training program. It involved weekly global conference calls, pilots, and feedback sessions from the various regional PMOs. The rollout was well received, and regional PMOs were eager to adapt and embrace project management.

Why were the outcomes for the two rollouts so different?

The answer has much to do with the different approaches taken by the respective teams. The team in the second scenario made sure that all the regional and local PMO representatives were involved in the rollout. They familiarized themselves with the local cultural, organizational, and procedural idiosyncrasies. They were responsive to local needs and maintained a careful balance between the need for standards and flexibility.

Cultivating a Studio or Lab Environment

Imagine if you could cultivate a culture like a young start-up where people are excited, they have skin in the game, and they want to come and play in the sandbox, not because they are forced to but because they want to. Of course, it is easier said than done; the question is, what should this sandbox look like and how do we define the boundaries to achieve the right balance? An example is to think of the idea of *Freedom with Fences,* an informal motto that has been made popular at Harley-Davidson. It enables employees to understand both the limits and the latitude they have to make improvements in their work processes. Project managers and teams are like artists who collaborate in a studio and treat each situation as unique and use personal judgment and creativity rather than relying on rote processes for every situation.

Typically, PMOs spend a lot of time building fences of rules, restrictions, forms, mandated processes, and methods as defense and control mechanisms, albeit with good intentions to achieve standardization and consistency. But these fences tend to go overboard and suffocate people and stifle judgment and creativity. Studios or sandboxes, by comparison, provide the platform for artistic freedom that can spark creativity and innovation. What if your project or PMO was more like a lab where you could experiment and learn?

Forced Compliance versus Voluntary Compliance and Commitment

In our experience in implementing project management practices and PMOs, we observe a clear contrast in the behavioral effects of the two opposing approaches of

Table 2.3: Forced Compliance versus Voluntary Compliance and Commitment

Forced Compliance	Voluntary Compliance and Commitment
Follow methods and processes to the letter	Follow methods and processes in the spirit
Minimal compliance with expectations	Exceed expectations
Aggravation and annoyance	Inspiring and energizing
Excuses and avoidance	Passion and commitment
Blame and finger pointing	Ownership and responsibility
Prescriptive answers and recipes	Insightful questions and inquiry
Restrictive and stifling	Enabling and creative
Hidden agendas, sabotage, passive aggressive	Uncompromising dialog, problem solving, and negotiation
Take advantage of established measures and metrics	Strive for meaningful measures and metrics
Controlling and constricting culture	Freedom and responsibility culture

forced compliance in rigid environments, versus voluntary compliance and commitment in responsive environments. Table 2.3 highlights the differences and provides a list of desirable characteristics to strive for.

HOW TO DESIGN EFFECTIVE BOUNDARIES

An oft-used cliché is that managing people or stakeholders is like herding sheep; the challenge is, how do you corral the sheep in such a way that you steer them in the direction you want them to go effortlessly? The answer lies in how you design the field to your advantage by focusing on cultivating a conducive environment and tilting it to your advantage, so people comply because they want to and not because they are forced to. While implementing project management practices, the fences of processes and methods should be developed and raised collaboratively as much as possible. They should be permeable and flexible with built-in mechanisms for feedback. And, of course, they should provide enough freedom for personal judgment, creativity, and innovation. The PMO should cultivate a culture where it is OK to *bend but not break* the rules to be agile.

Fair Process and Procedural Justice

People will respect the rules and boundaries of the fences if they believe it was a fair process. The idea of fair process and procedural justice is based on the work of two social scientists, John W. Thibaut and Laurence Walker, who combined their interest in the psychology of justice with the study of fair process. Focusing their attention on legal settings, they sought to understand what makes people trust a legal system so that they will comply with laws without being coerced into doing so. Their research established that people care as much about the fairness of the process through which an outcome is produced as they do about the outcome itself.

Fair process is based on three mutually reinforcing principles: the 3 E's—Engagement, Explanation, and Expectation clarity. We have added another element in our practice—Empathy.

Engagement means engaging stakeholders and proactively seeking their input, particularly in aspects of the project that will affect them most. Engagement provides a sense of confidence that their opinions have been considered.

Explanation details the decisions and makes sure the stakeholders understand the key points. You cannot assume that decisions are self-explanatory or straightforward. More importantly, explanation should also provide the background of why project decisions were made. This provides people with the context as they try to assimilate and adopt the changes from the project.

Expectation clarity describes the "new rules of the game." It requires clarification of the expectations and consequences brought about by the process.

Empathy involves putting yourself in the shoes of your stakeholders to understand and feeling the pain that the change is going to bring about. This helps you better plan to make the change process fair and just and connect with them from their perspective.

The PMO must seek procedural fairness by encouraging participative decision-making. Two-way communication helps dispel perceptual inequity. PMs who provide input into the decision-making process are more likely to support and implement procedures. Conversely, estrangement from the decision-making process can induce powerful resistance to change.

Communities and Collaboration

The idea of fair process needs a rich environment and appropriate culture to be successful. Communities of practice (CoPs) provide a fertile setting for these ideas to sprout and spread. How do you create a culture where people are excited about project management and PMO from the bottom-up and want to engage and contribute in a vibrant collaborative environment?

A community is defined by a shared interest or expertise. Practitioners collaborate to share and develop practices in their pursuit to improve project management and PMO practices. They become a community in the course of pursuing their interest, which leads them to participate in joint activities, share information, and help each other. In the process, they build relationships that further their efforts in the field. It is these relationships and the associated interactions that make the group a CoP. They share a common body of knowledge, resources, experience, and language (jargon) that enables them to learn from and contribute to the community. Thus, cultivating a collaborative culture and a sandbox that is responsive, but with the appropriate amount of rigor and discipline. The idea of communities will be further explored in Chapter 7.

Self-Regulating

Imagine if you are running late for a meeting and as you start to accelerate your car, your foot immediately hits the break, not because you see a policeman with a radar gun but because you encountered a speed indicator display (SID). A SID is a device that measures and displays the motorist's speed, to provide timely information to modify behavior to drive within the speed limit. A SID is essentially the friendly side of speed enforcement, nonthreatening but effective. Similarly, the PM processes need to be self-regulating and nonthreatening. For example, rather than setting a project review or escalation meeting, the PMO can design and implement preset triggers that provide timely feedback to self-regulate behavior in desired ways. Just as the SID is designed to slow traffic to a preset limit, PM processes can be used to define the boundaries with preset triggers for escalation. With growing capabilities to embed sensors everywhere, the ability to measure and provide real-time feedback and self-regulate behavior is increasing.

Scalable

One of the common complaints from project managers is that it takes them more time to complete the project documentation than the project itself. Even though it was a simple project, they had to apply all the steps to comply with the PMO methodology. To strive for global consistency and standards, a one-size-fits-all mentality sounds good but is not practical. Projects and programs by definition are unique with different characteristics requiring diverse approaches. Scalable processes and methods can be designed to address the unique aspects of projects. A simple project may need very limited process steps versus a complex project that may require more elaborate methodology steps.

Self-Eliminating

Good processes should have a built-in mechanism for changing or eliminating the process. We can all identify processes in our organizations that have survived way beyond their desired purpose. There are processes that are in practice and institutionalized simply because they have been done for a long time and nobody has questioned them. Part of the PMO governance should be a method to conduct periodic process reviews and decide when a process or practice is no longer useful or when it needs to be updated to make it useful again.

Desire Paths

Desire lines can usually be found as shortcuts where constructed pathways take a circuitous route. While shortcuts can be frustrating to landscape designers, some planners look to them as they map out and pave new official paths, letting users lead the way. Some educational institutions, including Virginia Tech and the University of California, Berkeley, waited to see which routes evolved organically as more people walked over these paths, before deciding where to pave additional pathways across their campuses. Similarly, we advise PMOs to sense and observe the existing desire paths of methods and processes, and adapt and reengineer PMO processes along end-user, customer, and stakeholder desire lines. Rather than an outside-in approach, this is an inside-out approach based on natural desires paths of customers and stakeholders.

SEVEN KEYS FOR EFFECTIVE PROJECT MANAGEMENT AND PMO FOCUS AND RESULTS

What are the themes across project management and PMOs that are thriving and perceived as valuable, delighting their stakeholders and making an impact? In our work with diverse organizations, from start-ups to established companies, government organizations, and nongovernmental organizations (NGOs) around the world, we have observed and codified seven themes that are common. Successful PMOs use the following questions as keys to unlock the right doors that lead to success and better focus and adoption of project management and PMO:

1. Who is our customer? How do we better focus on the customer and business?
2. How do we define success? How do we measure what matters?
3. How can we be better prepared to deal with the DANCE and disruption?
4. How can we cultivate a culture of connection, community, and collaboration?
5. How can we develop and enhance overall change intelligence?
6. How do we learn better and adapt? How do we develop signature practices that give us a competitive edge?
7. How can we simplify and build a Department of Simplicity?

Remember, if you are not using all of these keys as questions, some of the doors may not open. You are either not aware of it or not focusing on the right things. All of the seven keys are important for effective strategy execution. We find that if even one of these keys is missing, PMOs and strategy-execution offices (SEOs) can lose focus and implementation efforts can stumble.

Shaping the Future with Project Management and PMO

"By 2030, 25 percent of all transportation trips will be smart and driverless."
"By 2030, 25 percent of buildings will be 3D printed."

You might think that these bold statements are from one of the big Silicon Valley companies we hear about. Think again. These statements are from the government of

Table 2.4: How to Use the Seven Keys to Unlock Effective Project Management and PMO Focus and Results

Key	Door	Leads to ...	How?
Customer and Business Focus: Who is our customer? How do we better focus on the customer and business? How do we become a steward of value and impact?	Customer and Business focus	Alignment of strategy execution, greater customer focus, understanding of what drives business, ownership, and accountability for results and outcomes	WITGBRFDT?—Start everything with the question What Is the Good Business Reason for Doing This? Understand your business model canvas Develop a business plan for your PMO Address how you can help run, grow, and transform the business Covered in Chapter 4
Measurement: How do we define success? How do we measure what matters?	Measurement	Selecting the right metrics and continually evaluating performance and results	Define what success is? Develop a strategy execution scorecard PMO delight index Covered in Chapter 8
Adaptive: How can we be better prepared to deal with the DANCE and disruption?	Adaptiveness and agility	Speed; responsiveness; agility; rigor without rigidity	Not Either/Or, but AND; Ambidextrous Utilizing paradoxical frames; finding your sweet spot between rigidity and responsiveness Covered in this chapter
Connection and Collaboration: How can we break the barriers and cultivate a culture of connection, collaboration, speed, and agility?	Connection, community, and collaboration	Flatter structures, engagement, and excitement, ownership, and accountability, self-regulated governance, voluntary compliance, greater adoption	Cultivate a sandbox or studio; emulating the positive characteristics of lean start-up cultures Covered in Chapter 7
Change: How can we develop and enhance overall change intelligence?	Change-making	Greater change-readiness; increased change absorption and adoption capacity	Develop organizational, psychological, and emotional change intelligence Covered in Chapter 9

Table 2.4: (Continued)			
Key	**Door**	**Leads to …**	**How?**
Learning: How do we cultivate a learning environment? How do we learn better and adapt? How do we develop signature practices that give us a competitive edge?	Learning	Culture of failing fast, learning from failure, refrain from blind adoption of best practices; adoption of appropriate and useful practices	Rethinking best practices, cultivating signature practices Covered in Chapter 10
Simplicity: How can we simplify and build a Department of Simplicity?	Simplicity	Elegance and simplicity; greater buy-in, acceptance, and adaptability of processes, methods, and tools	Build a Department of Simplicity Covered in Chapter 11

Dubai, where they are constantly trying to find the sweet spot among the extremes. In my work and interactions with the PMO at the Roads and Traffic Authority (RTA) of Dubai, it has been an interesting journey to see how they try to balance and deliver on the bold goals, whether it is to fix the traffic problems in Dubai delivering on the goal of autonomous vehicles, or creating a roadmap for adopting 3D printing within the RTA.

I have seen the PMO evolve from a few project managers to an enterprise-level PMO, placed with the director general's office recognizing its importance and impact. With numerous mega projects at the cutting edge of technology, from autonomous vehicles to getting ready for future technologies like hyper-loop, it can be a challenge to find the balance between rigor and rigidity. As the director of the PMO shared, "It is not easy, but we have learned to focus on **what** needs to get accomplished, and shape the boundaries to get it accomplished, and leave the project teams to figure out the **how**."

With the push toward the idea of Dubai 10X, symbolizing experimental, out-of-the-box future-oriented exponential thinking in projects, they are exemplary in how they have learned to thrive on the edge in a DANCE-world. The way they strive for agility is by rating themselves on a future fitness indicator, which rates **foresight**—the depth of future envisioning opportunities, solutions, or threats and readiness with optimal processes, or rigid adherence to strategies and processes that threatens, versus **reaction speed**—measuring the reaction toward the opportunity or threat, as illustrated in Figure 2.4.

Project management and the PMO is having an impact not only within government but also in the region as they have sponsored an annual Dubai International Project Management Forum since 2014, and recently they instituted an award for innovation in project management.

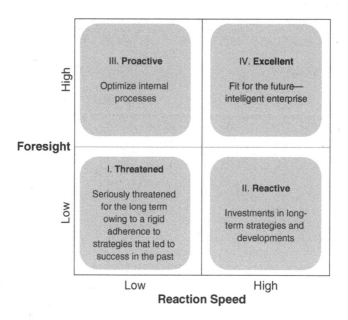

Figure 2.4: Future Fitness Matrix
Source: Adapted from RTA's Approach towards shaping the future.

How can you get project management and PMO fit for the future, with the right foresight and reaction time? That is what the rest of the book is about. The next chapter decodes the DNA of strategy execution.

KEY TAKEAWAYS

■ Traditional project management and PMO processes are about consistency and control and are designed to work well in a mechanical world, whereas the DANCE and disruption are characteristic of difference, failure, disorder, learning, and mutation.

■ Standards, processes, and structures can be beneficial, but the challenge is knowing how much is right for your organization and business. Failure stems from either too much or too little governance and process. The challenge is to thrive. You need to find your sweet spot between the two extremes.

■ The next generation depends on questions. In the past, we relied on answers, but in today's world as the DANCE intensifies and uncertainty increases, the value of questions goes up, and the value of answers goes down. Project/program managers and PMOs should learn to ask better questions and seek new perspectives to connect different aspects of the DNA of strategy execution.

■ On the one hand, there is a need to establish rigor with a sound governance structure of methods and processes. On the other hand, there is a demand for freedom and flexibility. This is indeed a primal paradox between the need for discipline and freedom at the same time.

- Next-generation project management and PMO is ambidextrous. It recognizes the importance of both the elements and does not focus on either/or. It acknowledges the tension between the paradoxical elements but understands that marrying the two reduces the undesirable effects of each and leads to new insights and blended solutions.

- True agility is achieved by aiming for rigor without rigidity. Balance foundational stability with dynamic agility. Strive to provide freedom with fences. Use ideas of fair process, community and collaboration, self-regulation, scalability, self-elimination, and desire paths to find and fine-tune your sweet spot. Cultivate a studio or lab environment where you can experiment, learn, grow and innovate.

3

DNA OF STRATEGY EXECUTION

Organizations are more like a living organism rather than a machine. A machine can breakdown, but organizational DNA can adapt and evolve to changing conditions and is resilient and can bend without breaking. The key is to decode the essential elements and find ways to strengthen them.

Leading Questions

- Why do we need the DNA of strategy execution?
- What are the elements of the DNA?
- What is the big deal about the DNA?
- How do you apply the DNA to your organization, project management office (PMO), and project management?
- What are the role, types, models, frameworks, and functions of a PMO?

For the past two-and-a-half years John has been leading a PMO in a leading energy company. The project portfolio is worth US$34 billion, spanning various energy generation, transportation, and distribution projects. His focus has primarily been on execution. One of the strategic priorities for the company is listed as execution and states that there is a need to focus on project management to safely deliver projects on time, on budget, and at the lowest practical cost while attaining the highest standards for safety, quality, customer satisfaction, and environmental and regulatory compliance. John has been busy with his team of about 30 working on various aspects of standards, controls, and processes. They feel pretty good about their accomplishments. They have established a standardized methodology and provided relevant training and documentation. They are getting 90 percent compliance to the PMO processes. However, as they have been tracking compliance with standard methods and processes, their project success rate has not gone up. To John's surprise, in a recent conference call with the business unit heads, six of out of eight executives said that the PMO was not necessarily adding any value to their business. What's amiss? What are John and his team not seeing? What can they do differently?

As the product backlog had increased and there was another series of missed dates and delays, Mary was asked to focus on reporting to provide objective status on how key initiatives and projects were progressing in this growing start-up. Mary was hired to bring a disciplined approach in this fast-paced, chaotic environment. In the past three months, she has managed to put in place a standardized reporting process and tool. Even though it has been a struggle, and it is a time-consuming effort each week to get the required data into the system, at least there is a consolidated dashboard. But the challenge is that now there is more time wasted in meetings discussing why the status is not accurate—the project managers do not think the data depicts the true status of the projects. The problem of delays and missed deadlines persists. What could Mary do differently?

With increasing number of requests and constantly being challenged about appropriate project prioritization by his business counterparts, the chief information officer charged the newly formed PMO to implement a portfolio management process for project selection and prioritization. Steve, the PMO director, researched portfolio management best practices and related tools and processes, and he attended workshops before he introduced a portfolio process. He was also successful in initiating a portfolio committee composed of various functional areas from business, information technology (IT), and finance. It has been a year and a half since this process was initiated and there is a published pipeline of projects. Initially, some of the stakeholders were excited that this would solve a lot of the issues and provide better prioritization and focus. The committee has met four times and deliberated and ranked proposed projects based on data put together by the PMO. However, there is increasing frustration as pet projects, and other initiatives seem to sneak-up on the list that were not part of the process. Over time, the portfolio process has come to be perceived as yet another PMO process without much value.

In each of the above examples, there seems to be the initial success, but it is short-lived or superficial. When you zoom out and look at the context, there is more to the story and raises key questions. ...

What could John, Mary, and Steve have done differently?

In the first scenario, John and his team are heads-down and comfortable in their domain of delivery and execution and are proud of their accomplishments, unaware of the wider business context and challenges. They are measuring and tracking standardization and certification and are not connecting to business results. They erroneously believe that standardization and process rigor will solve execution issues. They don't realize as our ongoing PMO survey since 2008 has revealed that a high degree of compliance (80 percent and above) to PMO processes does not correlate to project success. John should have questioned: Do we measure the right things? Are we addressing the business needs? What is the impact of standardization on project success and business results?

In Mary's case, she was able to implement the reporting tool, but she did not adequately question: Is the data necessary for the dashboard being captured and available? How can we engage project managers and involve them in the implementation to increase adoption? What are the lessons from past similar implementations? What are the necessary governance processes that need to be put in place for a successful implementation? What kind of behaviors are the new reports and dashboards going to drive?

The portfolio process seems fine on the surface, but the PMO director, Steve, should have questioned whether there is meaningful data available to rank the projects objectively. Are there established portfolio governance processes that are respected and followed? Are all the right stakeholders involved? What are the underlying politics and cultural issues that can sabotage a portfolio process?

In each case above, the common thread is that while they focused on solving their issue, they did not see the whole. They created a bubble around themselves that prevented them from seeing the whole and all the elements necessary that impacted their results. It is important to understand that everything is connected to everything else and the underlying interplay can have a ripple effect on results.

This is a typical pattern that we see repeated over and over in our practice that prompted the question: What if we could decode the DNA of effective strategy execution? Just as DNA contains the genetic instructions used in the development and functioning of all known living organisms, is there a code or blueprint containing the elements of management and strategy execution? If we can decode the DNA of management, we can identify the missing elements, the strengths and weaknesses, and disconnects. As discussed in Chapter 1, to thrive in a DANCE-world, organizations must recognize that their true nature is more like a living organism rather than a machine. A machine can break down, but a DNA can evolve and adapt and bend without breaking down. DNA is the instructional manual for life—from microbes to plants to human beings, it defines life, and a complete set of instructions are encoded in the organism's DNA. What are the elements of the DNA of strategy execution?

THE JOURNEY OF DECODING THE DNA

One of the core exercises we facilitate in our organizational project management and PMO practice is to pose the question, what is the purpose and functions of project management/PMO in your organization? We draw two circles on flip charts on either side of the room and record the responses. In 15 years of conducting this exercise, it follows a predictable pattern as seen in Figure 3.1. Initially, most of the responses belong to the right circle:

As we continue to prod more, we will gradually hear aspects that go in the left circle (Figure 3.2).

This is a revealing exercise and provides important insights. Typically, project managers and PMOs start-off and are comfortable on the right side, which is all about execution and getting things done. While they are heads-down focused on the what, how, and when of delivery, they may not be aware of the why. They work hard to deliver the outputs within time and budget, but they don't know what will be the outcomes of their actions and what will be the results and impact of their execution. They may rush to deliver a product or system, but the customer does not use it because it does not meet their needs. The PMO is busy standardizing methods and processes and training without connecting to the business reasons and objectives. The left

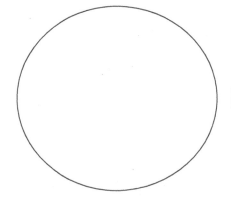

Projects on-time, on-budget, to-scope
Risk, quality, communications, human resources, procurement, integration, stakeholder management
Standardization (consistent methodology and processes)
Reporting & dashboards
Talent development – Training, coaching, mentoring, & supporting PMS

Figure 3.1: Identifying the DNA I

Focus on the right projects
Prioritization
Portfolio management
ROI
Profitability
Value
Impact
Benefits management
Customer-focus
Growth
Innovation

Projects on-time, on-budget, to-scope
Risk, quality, communications, human resources, procurement, integration, stakeholder management
Standardization (consistent methodology and processes)
Reporting & dashboards
Talent development –training, coaching, mentoring, & supporting PM

Figure 3.2: Identifying the DNA II

circle is focused on delivery and execution, and the right circle is about prioritization and making the right decisions to achieve benefits and value, which is the domain of strategy.

Strategy and execution are traditionally viewed as two distinct activities. It creates a disconnect between what an organization is trying to accomplish, and what is being done and executed on the ground on a day-to-day basis. This disconnect causes further breakdown across silos with increasing finger-pointing, and each blaming the other, for example: "Execution is the problem"; "We don't know how to get things done." On the other side, the people involved in execution blame the other side:" We really don't have a strategy"; "They don't know what they want"; "We don't have any priorities."

Unfortunately, the classic management literature has contributed to the gap by separating the thinkers from the doers and identifying strategy and execution as separate activities. In project management literature in recent years, there is a focus on distinguishing projects from programs and portfolios, with programs and portfolios focusing more on the strategic side. Although the intention is well-meaning, to bring the strategic focus in the organizational project management domain, the challenge is that instead of bringing together, it further fractures and separates the ownership and accountability of outcomes and impact (program focus), from outputs and deliverables (project focus). This approach can work in organizations that have a high degree of project maturity, but many organizations struggle to implement programs or separate the two. The question to consider is instead of causing cracks and separa-tion, how do we bring it together and encode ownership and accountability into the DNA at every level? Start-up cultures by nature have skin in the game and a greater degree of ownership of outcomes, their challenge is execution, whereas established companies have the opposite problem. The light bulb moment is to realize that you cannot separate strategy from execution, like the double-helix structure of the DNA, the two circles of strategy and execution are the core strands of the DNA that need to be woven together. The overlaps should feed off of each other, creating a virtuous cycle rather than isolated bubbles. Companies like Amazon strive to do exactly this. As Jeff Bezos of Amazon put it, "When you apply four things—customer focus, invention, investment, and operational excellence—they work together synergistically in all dif-ferent parts of our business." This is a good example of integrating the core elements of the DNA—the strategic elements of customer focus, investment, and invention, with the operational excellence of execution.

Take a look in the mirror and see which side you recognize yourself in—the left side of strategy or the right side of execution. Ask, where is my focus? Where is my organization's focus? Do we have a strategy? Is there clarity on our strategy? Is there a disconnect between strategy and execution? What are our strengths and weaknesses? What steps can I take to connect and align strategy with execution?

The right side of execution is all about performance—productivity, efficiency, and exploitation with the right people, processes, and tools (Figure 3.3). You can get better and improve performance, but performance by itself is not enough; it must be aligned with purpose to achieve results. Indra Nooyi of PepsiCo found a way to articulate and

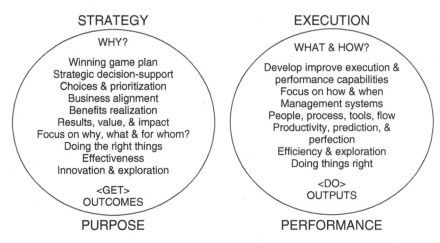

Figure 3.3: Identifying the DNA III

align strategy with execution and drive PepsiCo's sustainable long-term growth while leaving a positive imprint on society and environment with the theme of "performance with purpose" in 2006 when she took the helm as CEO.

Decoding the Other DNA Elements

To continue to identify the elements of the DNA, the next question is, how do we bridge the two sides? **Governance** is the strand that links strategy and execution. It lays the boundaries and decision-making mechanisms with a framework of policies, procedures, rules, guidelines, and oversight to steer and make sure there is alignment of actions on the execution side with the strategy to meet business goals and objectives. For example, you just completed a project phase and need to move on to the next; the governance mechanism is to follow the stage-gate process to ensure that the project still aligns with the business case, ensuring a link to the business side. In the PMO, you want to implement a project portfolio management (PPM) tool to improve execution; the governance mechanism might be to have a sound business case that justifies the investment and shows how the tool will benefit and impact strategy.

Connect is the binding element to identify disconnects and connect people, processes, and products to projects, programs, and portfolios across silos. You cannot accomplish much without the right relationships and connections. Connect aims to link stakeholders, silos, business priorities, and interfaces and interdependencies, by focusing on networks and connections, marketing communications, relationships, and community and collaboration. Connect detects systemic disconnects and bottlenecks, resolves communication and interface issues across organizational silos, and develops relationships and collaboration with stakeholders. Connect is the element that links the other strands of the DNA of strategy execution together. You can master the other strands, but success will hinge on whether the PMO has the right connections with the right parts of the organization, and knows how to communicate through them.

Measure is the feedback loop in the DNA that measures, collects, and provides relevant measurement information. Effective measure should provide feedback to learn and adjust performance in each area of the DNA. Measure aims to define and measure success. You can measure anything, but the challenge is to measure what matters; that drives the right behavior and performance. A strong measure strand should result in consolidated information and transparency with relevant reports, dashboards, and scorecards. It should provide measures of not just execution outputs but also strategy outcomes and, more importantly, focus on the leading measures that lead to the desired results before it is too late.

Change is the catalytic element in the DNA of strategy execution; it can either stimulate and drive action, or slow and stall your project or PMO. You can excel in all the other elements of the DNA, which provide the structure, process, and capabilities for strategy execution, but without change, you are jeopardizing adoption and intended outcomes of your initiative. Projects, programs, and PMOs succeed or fail based on how they understand, manage, lead, leverage, adapt, and adopt change from multiple perspectives—organizational as well as personal (psychological, emotional, neurological, and behavioral) change. Develop change intelligence by learning and adapting as you focus on change awareness, anticipation, absorption, and adoption of what you are trying to accomplish.

Learn must be ingrained into the DNA to learn, evolve, and mutate instead of a downward spiral of repeating the same mistakes over and over, without any improvement. All the other elements can be healthy and perform well, but learning provides an edge to evolve. A strong learn strand will help to harness the lessons learned; learn from failure; apply the lessons to make failure survivable, and continuously improve and innovate in each element of the DNA.

Context and customer-focus is the operating environment in which the DNA thrives. It is the nature of the business, purpose, vision and goals, the organizational structure, culture, and politics that determines how the DNA elements adapt and thrive, or are constrained and struggle. Context is like the operating system that provides the backdrop for strategy execution, and projects and PMO leaders must be mindful of their context and adapt and adjust their approach based on the business, culture, politics, and organizational dynamics. While you are doing all this, you must keep in mind that the DNA exists because of the customer, and customer experience and ultimately customer proliferation is the goal of the DNA.

The above elements have been identified over the past 15 years in our Next-Generation PMO and Portfolio Management seminars and practice working with a few thousand people in many organizations around the world. The challenge was to decode the DNA or identify the core elements of a successful PMO, which could be scalable and applied to any type of organization or business (Figure 3.4).

What Is the Big Deal about the DNA of Strategy Execution?

Each of the seven elements of the DNA is not new in itself; what's new is linking them together and understanding the interplay and impact of all the elements on each other and the overall organization and business. These elements can be found individually or in a combination of two or more, but rarely are all the elements linked

Context

Business - Purpose - Vision - Goals - Structure - Culture - Politics

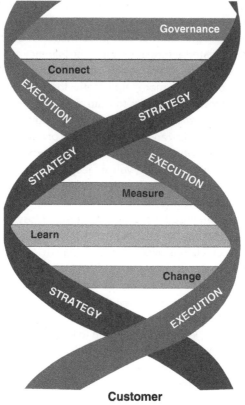

Customer

Creation - Focus - Experience - Impact

Figure 3.4: The DNA of Strategy Execution
Source: © J. Duggal. Projectize Group.

holistically in any management situation whether it is a project, program, PMO, or the overall organization.

As you analyze each of the scenarios at the beginning of the chapter, you will see an underlying pattern of the following cascading problems:

Disconnects. The factory setting of contemporary organizations is biased toward breakdown and specialization. Although the intention is well-meaning to make things more manageable, it creates vertical siloed processes and fragmented responsibility. There is an assumption that somebody is connecting the dots to link things horizontally, which does not always happen, resulting in disconnects and finger-pointing—disconnects between strategy and execution, between stakeholders, between what management intends and what teams assume, between customer needs and expectations, and more. Disconnects occur between governance processes and strategy goals, between execution and change adoption, between measurement and its impact on each of the other DNA areas, and more.

Lack of ownership. Disconnects result in lack of ownership. The strategy is not owned by the people who must implement it and vice versa. There is limited ownership, only for their own piece, and the end-to-end perspective is missing. There is no sense of overall skin-in-the-game, causing a lack of ownership of outcomes and impact. If you are on a plane and you see some smoke, you are not going to think, I am sure somebody else is seeing it, and they will let somebody know. Let me just stay focused; I need to get to my destination on time. Of course, you don't behave like that. You are going to let somebody know immediately. You know your life is on the line and you are clear about the outcome, and you take ownership without even thinking. Survival is part of your DNA. Ironically, lack of ownership is pervasive in organizations.

Lack of holistic perspective. There is a whirlwind of activity in each area, and individual silos are busy dealing with their issues and putting out their own fires. This creates a vicious downward cycle in which they are blinded by the limited view of their silo. The big picture perspective to zoom out and look at the "whole" is missing. There is a lack of systems perspective on how everything is connected to everything else and has a ripple effect. It is not surprising that many project managers and PMOs are tilted toward execution and delivery and are comfortable in that circle. They don't connect adequately to the other elements of the DNA and don't grasp the big picture business perspective and how the PMO can add value where it is needed. By looking at the whole, you can spot the disconnects and imbalance.

Lack of agility. Disconnects compounded with lack of ownership, and holistic perspective make the organization slow and stodgy and unresponsive to the changing landscape. Layers of management in multiple silos creates bureaucracy, which hinders agility and flexibility. The organization is busy trying to stay afloat in the whirlwind of day-to-day issues and not prepared to deal with the DANCE and disruption and adapt and steer itself in a new direction.

PMOs typically get started with a particular focus to develop execution capabilities, to establish governance, or to provide reports. While the PMOs deal with these issues initially, they only provide limited value and cannot make a greater impact without a holistic approach. To address these issues, they need to look at the DNA.

Just like understanding of the DNA changed the approach to medicine, the DNA of strategy execution will change your approach to management. Decoding the DNA can provide vital information on how the system works and how the elements are replicated at every level, like the DNA in every cell. All the elements are part of on interdependent and integrated whole. Every element matters because it impacts others. Even if one of the elements is weak or missing, it impacts the health and proper functioning of the whole. A healthy DNA can adapt and evolve to changing conditions and is resilient and can bend without breaking. Just as each cell contains the DNA defining who we are, the DNA of strategy execution replicates at the project, program, portfolio, PMO, operational, and organizational level. Like the instructional manual for life, the DNA of strategy execution is subtle, elegant, and complex.

You may start with focusing on one area, but keep in mind the whole while working on the individual elements of the DNA is important. You may not be able to address or solve all the underlying issues but starting to see the whole goes a long way. For example, in Mary's case above, even though her mandate is to provide reports and dashboards and focus on **measure**, she needs to zoomout and make sure she measures what matters from both sides, **execution,** and delivery, as well as **strategy** metrics. She needs to make sure there are appropriate **governance** processes for measurement. She needs to **connect**, collaborate, and build relationships with the right stakeholders and engage them with appropriate marcom strategies as she connects the dependencies across different areas. Mary has to **learn** and adapt as she practices **change** management in the current **context** of her environment while making sure her efforts are ultimately **customer-focused.**

Imagine you could encode these elements into the DNA of your organization at every level. Of course, it is not going to be easy. One of the successful entrepreneurs, Danny Meyers, has cultivated a culture that is embedded with the mantra of "ABCD—always be connecting dots," in his successful top-rated restaurants, which serve over 100,000 customers daily. Everyone from the greeters to the servers is trained in ABCD, and everyone owns the customer experience and outcomes. The DNA elements by nature help you to connect the dots across the board and embed ownership at every level. Instead of a downward spiral, the DNA creates a learning loop of adaptation as the DNA strengthens with experimentation and iteration. Imagine, every project/program manager, PMO, or executive linking, balancing, and optimizing to each of the DNA elements as they address any particular issue. It is OK to have separate teams working on their individual element; the challenge is to train them to connect to other areas and embed shared responsibility for ownership of business outcomes.

A PMO in a banking environment shared an example of how, as they applied the DNA framework, one of the teams involved in an ATM upgrade project asked a question about how this was a temporary fix and would be unnecessary based on what they were working on for the next release. This saved $22 million for the bank. Previously, they would have followed what they were told to do and would have completed the upgrade, assuming somebody else was worried about why we need to do this, without asking any questions.

THE DNA UNDER THE MICROSCOPE

As you look under the microscope, you can identify the strands in each of the elements of the DNA in Figure 3.5. Each of the elements will be explored in detail in the following chapters.

Developing DNA Intelligence

Another aspect we are going to look at is how to improve, mature, and develop intelligence in each of the DNA areas. Table 3.1 lays out the foundational questions that will also be explored in subsequent chapters.

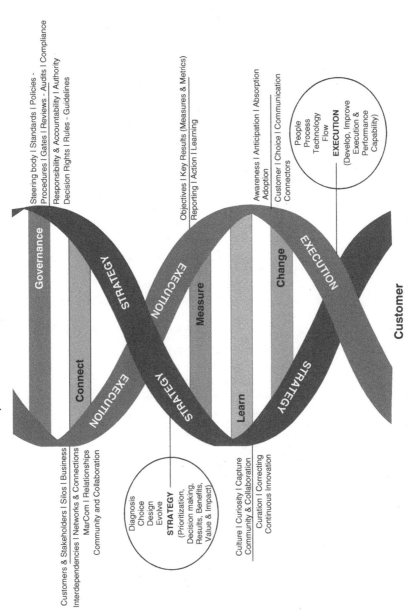

Context

Business - Purpose - Vision - Goals - Structure - Culture - Politics

Steering body | Standards | Policies -
Procedures | Gates | Reviews - Audits | Compliance
Responsibility & Accountability | Authority
Decision Rights | Rules - Guidelines

Objectives | Key Results (Measures & Metrics)
Reporting | Action | Learning

Awareness | Anticipation | Absorption
Adoption

Customer | Choice | Communication
Connectors

People
Process
Technology
Flow
EXECUTION
(Develop, Improve
Execution &
Performance
Capability)

Governance

Connect

Measure

Change

Learn

STRATEGY

EXECUTION

EXECUTION

EXECUTION

STRATEGY

STRATEGY

Customers & Stakeholders | Silos | Business
Interdependencies | Networks & Connections
MarCom | Relationships
Community and Collaboration

Diagnosis
Choice
Design
Evolve
STRATEGY
(Prioritization,
Decision making,
Results, Benefits,
Value & Impact)

Culture | Curiosity | Capture
Community & Collaboration
Curation | Correcting
Continuous Innovation

Customer

Creation - Focus - Experience - Impact

Figure 3.5   The DNA of Strategy Execution under the Microscope

Source: © J. Duggal. Projectize Group.

Table 3.1: DNA Strategy Execution Questions Framework	
DNA Element	**Questions**
Strategy	Is there clarity on our strategy?
	Are we making the right choices and focusing on the right things? How can we prioritize and make better decisions?
	Do our projects and activities align with our business, mission, vision, and goals?
	How can we improve our results, value, and impact?
	How can we do better exploration, innovation, and growth?
Execution	How do we assess, develop, and improve execution capabilities?
	How can we reduce firefighting and heroics?
	What people, process, and pipes do we need to have in place?
	How can we develop capabilities with the right talent and skills?
	Do we have the right combination of methods, processes, and systems capabilities?
	What are the pipes that connect people and processes—the inputs and interfaces that enable or impede execution?
Governance	Do we have the appropriate decision-making governance mechanisms—structure, roles, processes, roles, responsibility, authority, and accountability—for our current environment?
	How can we design governance for greater ownership and accountability of outcomes?
	How do we implement effective governance in an agile DANCE-world?
Measure	Are we measuring the right things?
	How do we define success?
	How can we measure what matters?
	Are our measures driving the right behaviors?
	How can we measure and demonstrate value and impact?
Connect	How can we reduce disconnects?
	How can we bridge and connect across silos?
	How can we connect and build relationships with stakeholders?
	How can we identify and resolve linkages and dependencies?
	How can we build community and foster collaboration?
Learn	How do we cultivate a learning culture?
	Are we learning from our success and failures?
	How can we fail fast and make failure survivable? How can we avoid repeating mistakes?
	How to curate and disseminate institutional knowledge and lessons learned?
Change	Are we change-ready?
	Are we change enablers or change blockers?
	How to catalyze change action?
	How to anticipate change?
	How can we get better at adoption?
	How can we develop change intelligence?

As the DNA has evolved over the past 10 years and we practice and apply it in many organizations, a common observation is that this is the DNA of management. You can use the questions in Table 3.1 to apply the DNA at the organizational, PMO, project, program, or portfolio level. You can also apply the DNA to classic management functions like finance, marketing, human resources, and so on. For each of the areas, start with the questions from each of the seven DNA areas and diagnose the strengths, weaknesses, and pain points. Think about it—for anything you are doing, you have to know what you are doing and how to get it done (**execution**); you have to know why you are doing it and make sure you are making the right choices (**strategy**); you need a sound **governance** process and methods, you need to monitor and track with **measurement**, you need to **learn** from your failures, as well as successes, and **connect** with the right stakeholders and customers, and you need to adapt and adjust to appropriate **change** and adoption. DNA is foundational and methodology agnostic whether it is traditional or agile. By nature, it is designed for agility, adaptation, and mutation depending on the evolving context.

GETTING STARTED: HOW TO APPLY THE DNA TO PROJECT MANAGEMENT OR PMO

Diagnosing the Pain

One of the questions that I love to ask is, why do you need a PMO?

One of the common questions that people not familiar with PMOs may pose is, why do we need a PMO?

I am not surprised to hear "standardization," "consistency," "improved capabilities," "better alignment," "return on investment," and so on. The problem with these responses is that they all sound like vitamins—they are too generic and nice to have clichés that could be true for almost any initiative. Executives and stakeholders have heard them before and are not compelled to prompt action. Before vitamins, what they want is painkillers. It is important to understand this distinction, as we see this mistake repeated over and over, where PMOs are selling vitamins instead of addressing the pain. Like a good doctor, you must diagnose before you prescribe; otherwise, it is malpractice.

You can start by identifying the symptoms of organizational pain. Do we have too many projects and too few completed? Are "flavor of the month" initiatives becoming a steady diet? Are our best performers demotivated? Is it impossible to track key project? Is too much focus on next-generation products forcing daily operations out of control? Or highlight the cycle of organizational firefighting. There isn't enough time to solve all the problems; solutions are incomplete; problems recur and cascade; urgency supersedes importance; many problems become crises; performance drops.

Start by assessing the PMO from the perspective of each of the seven DNA elements and evaluate your strengths, weaknesses, and missing links. The current point of pain and business needs of the organization will determine the priority of focus.

Focus on one or two issues; research and analyze them and get specific data to highlight the pain and build a convincing story. The key is to diagnose correctly and pick the pain that impacts the business and the ones that executives and stakeholders care about. Identify which element of the DNA is the source of the pain and how it connects and overlaps with the other areas of the DNA.

Melissa was hired as a senior project manager to bring about some order to the crazy world of projects in a gaming company in Las Vegas. For the past 18 months, she has made some progress but thinks that to make an impact they need a PMO with a supporting PPM tool. She has pitched the idea promising standardization and consistency, but she is not getting much traction; management wants her to focus and get the projects in control. As she learned and applied the DNA of strategy execution to the PMO, she changed her approach to first diagnose the pain from a business and organizational standpoint. The revenue recognition for this gaming company was directly tied to how many system installs they completed. Melissa did the research and created a road map of all systems installs based on estimated recognition for one to three quarters (**strategy/business element**). She specifically highlighted how this pain was directly impacting revenue in the next three quarters in the stream of $28 million. Next, she highlighted how currently there is no good way to tell if we have enough resources to do the work we are scheduled to complete in the time frame we are being asked to complete from the **execution** side. She demonstrated how a PMO tool with a built-in **governance** mechanism would provide visibility to resources and capacity planning. We currently have no good way to tell if the amount of money the company charges for work is too much or too little, as we have no way to **measure** cost/time. She showed them how the right metrics could be used to provide feedback for adjusting sales pricing decisions. She did this by **connecting** and building relationships one-on-one with the key stakeholders, identifying their specific pains and then linking it to the overall business and organizational issues. The key was to research, analyze, quantify, and be specific regarding the business issues and link it to their specific pain. Also, the linkage of all elements of the DNA is crucial.

Stakeholder Empathy Map

Besides the organizational pain, you also should identify the key stakeholders and customers and identify their pain. The stakeholder empathy map (SEM) is a powerful tool developed by visual thinking company, Xplane, that can be used to gain deeper insights into your stakeholder's perspective. SEM helps you to step into the shoes of your stakeholders, users, and customers. It helps you to uncover pain as well as potential gains from the PMO and meet unmet needs as you design the PMO from

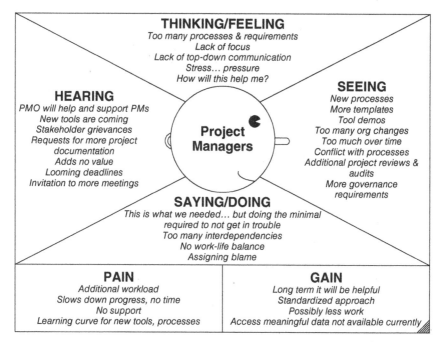

THINKING/FEELING
Too many processes & requirements
Lack of focus
Lack of top-down communication
Stress… pressure
How will this help me?

SEEING
New processes
More templates
Tool demos
Too many org changes
Too much over time
Conflict with processes
Additional project reviews & audits
More governance requirements

HEARING
PMO will help and support PMs
New tools are coming
Stakeholder grievances
Requests for more project documentation
Adds no value
Looming deadlines
Invitation to more meetings

Project Managers

SAYING/DOING
This is what we needed… but doing the minimal required to not get in trouble
Too many interdependencies
No work-life balance
Assigning blame

PAIN
Additional workload
Slows down progress, no time
No support
Learning curve for new tools, processes

GAIN
Long term it will be helpful
Standardized approach
Possibly less work
Access meaningful data not available currently

Figure 3.6: Project Managers Empathy Map

your stakeholder's perspective. Refer to samples of empathy maps from some of the key stakeholders of the PMO: project managers (Figure 3.6) and executives (Figure 3.7). Similarly, you can create maps for the PMO, functional managers, and other key stakeholders.

The SEM can be used in different ways. The PMO can do the empathy mapping exercise itself based on their observations and assumptions. Depending on their relationship with the stakeholders, they can validate it later with the stakeholders, or they could work on it together with them.

Identifying the Customer?

A powerful exercise to conduct with the PMO team is to ask the question, who is our customer? Is it the team we are handing over the deliverable to? Is it the sponsor? Is it our executives? Is it the end users? Is it the end customer of the business who buys our product or service?

I have facilitated this exercise in many organizations, and the typical responses are executives, scrum teams, team leads, functional managers, project managers, product owners, marketing/sales, and end users. They often forget the most important stakeholder. Only after prodding them and asking if they were missing any other important stakeholders does the light bulb go on and they identify the end "customer."

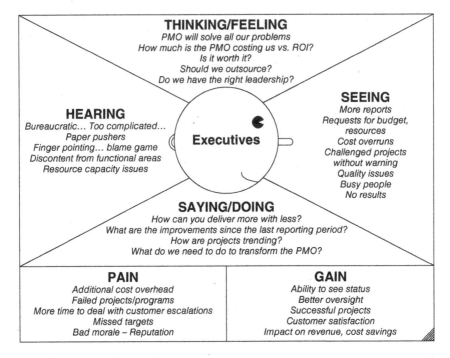

Figure 3.7: Executives Empathy Map

One of the themes of this book is that the customer is at the center of everything and customer focus and customer success should be the priority for project managers and PMOs. Ultimately, project managers and the PMO should focus on how the customer can succeed with the deliverable or service of the projects and programs. As you are working on any aspect of the DNA, always think about *customer-experience* or *customer-success,* which will lead to not only project/program or PMO success but overall strategy-execution success.

WHAT ARE THE ROLES, TYPES, MODELS, FRAMEWORKS, AND FUNCTIONS OF A PMO?

PMOs have been around for a while and have become a common fixture in many organizations. Despite this, the core purpose and function of a PMO continue to be questioned and debated. The situation gets further convoluted in organizations in which PMOs proliferate at multiple levels with overlapping or conflicting functions. Sometimes when there is clarity of purpose, however, having too narrow a focus limits PMO value.

Defining Your PMO

According to the *Project Management Body of Knowledge (PMBoK) Guide*, 6th edition (2017), a project management office (PMO) is

> an organizational structure that standardizes the project-related governance processes and facilitates the sharing of resources, methodologies, tools, and techniques. The responsibilities of a PMO can range from providing project management support functions to the direct management of one or more projects.

A PMO is an **organizational office** to organize, facilitate, and enable the effective management, execution, and results of projects, programs, or portfolio of projects. As a PMO evolves, it can become an **organizational approach** to organize, facilitate, and enable the effective implementation of organizational project management (OPM)—the systematic management of projects, programs, and portfolios in alignment with the achievement of strategic goals.

Besides organizational PMOs, there are also project or program offices responsible for the business and technical management of a specific contract or project/program. Often, strategic initiatives or large enterprise programs like a transformation program or global rollout of a system will have a program office, also known as a PMO.

The PMI Program Management Standard, 4th edition (2017), defines a program management office as

> a management structure that standardizes the program-related governance processes and facilitates the sharing of resources, methodologies, tools, and techniques. A program management office often also supports training and facilitates the sharing of resources, services, methodologies, tools, and techniques. Program management offices may be established within an individual program to provide support specific to that program, and/or independent of an individual program to provide support to one or more of an organization's programs office is an important element of the program's infrastructure and an aid to the program manager. It may support the program manager with the management of multiple projects.

You must work with your stakeholders and design your own approach in how you define the purpose, role, and functions of your PMO. The specific form, function, and structure of a PMO for your organization will depend on factors, such as:

Your current context
Your pain points, challenges, and needs
The size and type of your organization
The nature and scope of projects
Your organization's project management maturity

PMO Functions and Activities Service Catalog

Refer to Appendix B, PMO Functions and Activities Service Catalog, which is a detailed listing of the possible PMO functions, activities, and services, organized by each of the DNA elements.

Developing a Mission/Purpose for Your PMO

Remember John from the beginning of the chapter who was leading the PMO in an energy company? One of the questions I asked John was, do you have a mission or purpose for the PMO? He proudly sent me a link to their PMO website and their mission:

We add value by:

- Creation and support of the key project management processes and tools
    - Cost, Schedule, Change, Gating and Risk Controls
    - Clarity on "must-be-achieved" policies and principles
    - Subject Matter Experts to advise on unique project challenges
- Assisting with initial project setup, control staff training, definition of standard reporting
- Conducting process health checks to ensure achievement of mandated controls
- Assist projects with hiring roles, competency, and Rolodex of controls staff

As you review this mission, what comes to mind? Too many words ... primarily focused on execution and governance issues ... top-down controlling language— "must-be-achieved policies," "mandated controls," ... uninspiring and unengaging ... inside-out project management and PMO perspective, rather than outside-in, the business perspective, and what the customer needs.

We worked with their PMO team and facilitated a workshop. As they became familiar with the DNA, light bulbs started to go on, and they realized that they had to reinvent their mission and clarify their purpose to apply the missing elements of the DNA. After a lot of questioning, reflection, discussion, and validating assumptions, they came up with the following reinvented mission:

PMO will enable a collaborative approach to delivering projects that add business value in support of our growth strategy, by promoting the "(Company) Way" of executing projects through:

- Consistency
- Flexibility
- Simplicity

The PMO will provide and support the following value-added functions/services:

1. Project Management processes, tools, and training (Cost, Schedule, Change, Gating, and Risk Controls)
2. Identify opportunities for delivering business value
3. Enable project performance and reporting
4. Foster (champion) cross-project communication, relationships, and knowledge transfer
5. Project resource capacity planning and coordination

They found a way to apply the DNA, think holistically, and link aspects of strategy and business, change, learn, and connect that were missing from the PMO. The idea is not that the above mission is perfect and you can cut and paste to transform your PMO. It was just right for this organization at this time. For you, it may be different based on your current context and pain points.

Another PMO that was getting started in the media business applied the DNA and came up with this mission.

> Provide simple yet effective project execution and resource management support for strategic decision making to deliver quality, customer value, and enable business success.

After they had this mission for about a year, they were still struggling to get traction with the PMO. They realized that this mission was still driving project execution and resource management support even though it had all the right words. To reinvent themselves, they simplified the mission:

> Facilitate, enable, and connect strategic execution of project portfolio for effective decision making, customer experience, and business success.

This exercise was one of the key drivers that transformed the perception of the purpose and expectations of the PMO both from an internal PMO perspective, as well stakeholder and customer expectations.

Another PMO in a global retail business had been struggling to make a mark and demonstrate their relevancy and value. They conducted an exercise to identify organizational pain points and see how the PMO could do better. Recurring themes were: too much work, too much process, too complicated, lack of connection, lack of priorities, and too many dependencies.

After a PMO facilitation exercise, they came up with a short and elegant statement, "Simplify and connect to drive results." This transformed their approach and perception, and after five months the PMO was elevated to the enterprise level and became a go-to organization for driving strategy execution.

PMO Frameworks

The Project Management Institute's (PMI) *PMBoK Guide* defines three types of PMO:

1. **Supportive PMO.** "Low control," providing support and coaching to project managers and project teams with tools, templates, information, technical support, and training.
2. **Controlling PMO.** "Moderate control," governance focus by implementing standard methods and processes and monitoring compliance.
3. **Directive PMO.** "High control," direct responsibility for managing projects/ programs and direct accountability for their success.

Gartner Group, the IT advisory service, outlines four types of PMOs:

1. **Activist PMO**—taking an activist and enabling, not controlling, approach.
2. **Delivery PMO**—focusing on delivery and execution.
3. **Compliance PMO**—emphasis on governance and compliance.
4. **Centralized PMO**—centralized standardization, coordination, and tracking at an enterprise level.

PMOs can be implemented in a combination of various roles, personas, and functions depending on the organizational context, structure, politics, culture, leadership, and PM focus and understanding.

Table 3.2 lists the type of PMO or model, the role, persona, function, and the primary DNA element focus incorporating a variety of models we have encountered based on our experience.

What Type of PMO Approach Is Right for You?

As we discussed earlier, your PMO approach will depend on your organizational context, structure, politics, culture, leadership, PM focus, and the understanding and intention of who is involved and sponsors the PMO. At a broad level, these aspects steer the PMO to adopt one of three approaches—a **service, controlling, or partner** mentality. Also, whom the PMO reports to will define its function and mentality. If it is sponsored by finance, typically the PMO focus will be on measure (emphasis on reporting, cost, budget); if it is departmental or functional, the focus will be on execution and delivery. At the enterprise level, it is more strategic.

Start with one area that is a pain point or organizational priority, focus on it, and show results. These results can be leveraged to connect and expand influence to other areas of the DNA. Initially, don't tread openly where you don't belong. You don't have to show the DNA model and talk about it. It may intimidate some stakeholders. The secret is that as you focus on one pain area, you are finding ways to connect and add value to the other areas of the DNA.

PMOs exist at different levels within the organization, and one organization can have multiple types of PMOs. Table 3.3 lists types of PMOs at different levels, based on a survey of organizations in PMI's *Pulse of the Profession* report on PMOs.

The specific form, structure, and staffing of the PMO will vary depending on the function, level, and reporting relationship. Centralized PMOs, where project managers report into the PMO, versus decentralized PMOs, where project managers report to functional areas, are approximately divided 46 percent to 54 percent according to the Projectize Group ongoing survey since 2008. Both have their pros and cons. If the PMO is a directive PMO responsible for the success of the projects, centralized PMOs can give more clout to the PMO. If the PMO is a support-oriented PMO, it does not matter as much.

Table 3.2: PMO Models, Roles, Personas, and Primary Functions

DNA Focus Area	PMO Model	Role	Persona	Primary Function
Strategy	Strategy Execution Office (SEO) Office of Strategy Management (OSM) Portfolio Management Office (PfMO) Results Management Office (RMO)	Strategic decision-support	Thinker Designer Architect	Decision-support Portfolio prioritization and balancing Strategy execution
Execution	Project Support Office (PSO); Project/Program Management Office (PMO)	Delivery Service and support Coaching and mentoring Directive	Doer Executor Coach Mentor Chief of staff Catalyst Facilitator Expeditor Enabler	Managing, supporting, enabling, facilitating, expediting Coaching, mentoring— Projects and Programs
Governance	Project/Program Management Office (PMO)	Standardization Compliance Review and audit	Auditor Process police Control tower	Establish and implement standards, methodologies, policies, procedure, rules, guidelines Oversight
Connect/ Learn	Community of practice (CoP) Center of excellence (CoE)	Integration Coordination Collaboration	Conductor Curator Partner	Coordination Linking and Connecting Filtering Bridge building Curating
Measure	Performance Management Results Management Office (RMO)	Reporting and analysis	Weather-station Reporter Clean room	Reporting, data analytics, key performance indicators (KPI), Objectives and key results (OKR), filtering, dashboards, scorecards
Change	Transformation Office Organization Change Management Office (CMO)	Change management	Change agent Changemaker	Organization change management

Table 3.3: PMO Frameworks	
Organizational / Business Unit / Divisional / Departmental	54%
Project specific / Project Office / Program Office	31%
Project Support / Services / Controls office	44%
Enterprise / Organization-wide / Corporate / Portfolio / Global PMO	39%
Center of excellence / Center of Competency	35%

Strategy Execution Office

As we have discussed in this chapter, there is no one best approach to the type of PMO you can implement. Your approach will depend on your context. Regardless of what you call it or how you start, your goal should be to connect and link the different aspects of the DNA. If you are successful in bridging the gap between strategy and execution and linking the seven elements of the DNA, you will evolve toward a true strategy-execution office (SEO). The SEO is a holistic and hybrid approach, constantly adjusting with the aim toward balanced strategy execution and results.

It would be ideal to have a strategy-execution office that connects all the elements holistically, with a balanced and aligned portfolio, but that would be more like organizational nirvana. You can take different paths and combinations of roles, personas, and functions to evolve toward a type of SEO, regardless of what you call it. You can get carried away with the name or what you call the PMO. The name or even the word *office* can scare people and have negative connotations. What's more important is how you apply the DNA and practice the idea of holistic strategy execution with or without the name.

DEVELOPING DNA MATURITY AND INTELLIGENCE

Typically, organizational project management and PMOs focus on maturity—how to mature project management capabilities. Over the past few decades, there have been many maturity models starting with systems and IT and spreading into other industries. The various maturity models including PMI's Organizational Project Management Maturity Model (OPM3) have met varying degrees of limited adoption and success. The lesson learned from various maturity model implementation is that, although it sounds like a common-sense progression, each organization is unique with its own context, culture, and politics and it is hard to implement and more importantly adopt and practice by-the-book aspirational best practices in the real-world complexity of today's DANCE-world. You can have all the check list practices, but it does not mean they will work in the ever-changing DANCE context. In other words, you can be mature, but that does not mean you are smart and agile to survive the DANCE and disruption.

The strategy-execution continuum provides a road map to plan the evolution of the PMO. We have applied this continuum in many organizations in different industries around the world. It generates a lot of discussion among PMO stakeholders and provides a focus on a timeline to evolve the PMO. The original four stages of maturity models are based on **Standardize, Measure, Control, and Improve.** For today's DANCE-world, we have evolved these stages to **Standardize, Measure, Learn, and Innovate.** Learning and innovation are what truly sustain organizations, not merely impeccable control and improvement of the same processes. This continuum aims to bridge standardization and measure, which are about consistency and control, with learn and innovate, which are about capitalizing on the difference, failure, disorder, learning, and mutation. The strategy-execution continuum is mapped in Figure 3.8.

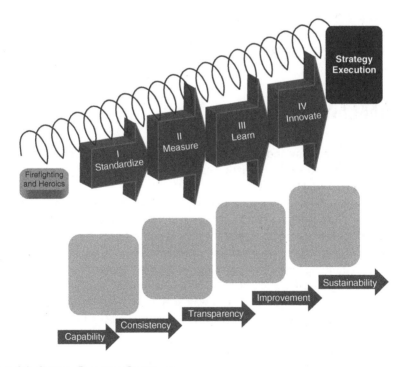

Figure 3.8: Strategy-Execution Continuum
Source: © J. Duggal. Projectize Group.

Standardize. Developing capabilities by focusing on standardization for consistency and repeatability.

Measure. For visibility and transparency to drive performance and results.

Learn. To improve predictability, while developing experience and agility to evolve.

Innovate. For better customer experience and customer creation, and sustained value generation and impact.

You can map each of the stages on a timeline and list what you want to focus on in each of the boxes underneath. For example, initially the PMO's focus in the first six months might be on standardization, and you might list, develop, and implement a standard methodology, provide training, and so on.

This is one aspect as you plan your road map along the strategy-execution continuum. Another aspect is to develop intelligence in each of the areas of the DNA. Use the DNA Strategy Execution Questions Framework to assess and develop intelligence in each of the DNA elements. Each of the following chapters will provide a list of questions to assess and develop strategy-execution intelligence.

"If you have individual actions without strategy it is random, if you have strategy without individual actions it is fantasy." - Dave Ulrich

Starting with the next chapter, we will put the DNA under the microscope and look into the details of strategy and then execution, followed by the other strands, to connect individual actions and organizational strategy execution in a holistic and coherent way.

KEY TAKEAWAYS

- Just like DNA strategy and execution are the two foundational strands that need to be woven together. The overlaps feed off of each other, creating a virtuous cycle rather than isolated bubbles.
- You can get better and improve performance, but performance by itself is not enough. It must be aligned with purpose to achieve results.
- Each of the seven elements of the DNA—*Strategy, Execution, Governance, Connect, Measure, Change, Learn*—is not new in itself; what's new is linking them together and understanding the interplay and impact of all the elements on each other and the overall organization and business. The DNA mindset helps to connect the broken strands back together in a holistic way, keeping in mind the *Context* (Business, Purpose, Vision, Goals, Structure, Culture, & Politics) and the *Customer* (Creation, Focus, Experience, & Impact).
- It is important to understand the distinction between selling vitamins versus painkillers. You must diagnose the pain before you prescribe project management and PMO solutions; otherwise, you are committing malpractice.
- Start by assessing the PMO from the perspective of each of the seven DNA elements and evaluate your strengths, weaknesses, and missing links. The current point of pain and business needs of the organization will determine the priority.

■ As you are working on any aspect of the DNA, always be obsessed with the customer and think about *customer-experience* or *customer-success*, which will lead to not only project/program or PMO success but overall strategy-execution success.

■ Your PMO approach should depend on your organizational context, structure, politics, culture, leadership, PM focus, and the understanding and intention of who is involved and sponsors the PMO.

■ If you are successful in bridging the gap between strategy and execution and linking the seven elements of the DNA, you will evolve toward a true strategy-execution office (SEO). The SEO is a holistic and hybrid approach, constantly adjusting and balancing with the aim toward true strategy execution and results.

■ The strategy-execution continuum provides a road map to plan the evolution of the PMO. Learning and innovation are what truly sustain organizations, not merely impeccable control and improvement of the same processes.

■ Use the DNA Strategy Execution Questions Framework to assess and develop intelligence in each of the DNA elements.

4
STRATEGY

Strategy without tactics is the slowest route to victory. Tactics without strategy is the noise before defeat.

Sun Tzu, Chinese Military Strategist, c. 490 B.C.

Leading Questions

- What is strategy? What it is not?
- What are the core strands of strategy?
- What is the role of project management (PM) and project management office (PMO) in strategy?
- How do we clarify, connect, and catalyze a coherent strategy across the board in everything we do?
- How do we select, prioritize, and balance projects/programs with portfolio management?
- How do we track the benefits and value of our initiatives and projects?
- How do we develop strategy intelligence?

What if you are a leading energy company that practically invented the market in your segment. For the past 25 years you did not have much competition, but now you have to deal with disruption in your industry. You unexpectedly lose a bid that was almost a guaranteed win. Up to this point, the focus was on execution, building technical capabilities and competence. The projects organization was proud of its accomplishments and timely execution of technically challenging projects. Now, suddenly, you are inundated with the reality of declining costs, reducing margins, increased competition, oversupply in the value chain, and changes in the underlying business model of the industry. One of the executives of this European energy company questions, "Up to now we have been measured on execution and delivering high-profit-margin projects on time, and we have not faced any competition. What do we do now? How do we win in this disruptive environment?"

In the convulsive changing world of the U.S healthcare industry, a PMO had survived on the sidelines by providing execution support for 10 years. After a new director took over and struggled to show value of the PMO, she had a transformative aha moment in my next-generation PMO seminar. She went on to transform and grow the PMO from a portfolio of 5 projects with a staff of 7 to managing a US$4 billion portfolio of 38 projects with a team of 60. The PMO became a focal point, and every proposal highlighted the PMO's execution capabilities as a core competency. What was the turning point? What changed the perception of the PMO from a sideline execution support entity to a game-changing influencer and strategic partner?

One of the largest banks in Asia was struggling with too many initiatives and projects and lack of prioritization. It decided to implement a portfolio management process. It formed a portfolio committee that, after much deliberation, came up with a project classification and ranking criteria based on classic financial risk-reward scoring for each project. Now the bank had a good inventory, and the projects were ranked on a prioritized list. This was a little better, but not much had changed. It was hard to make tough decisions to terminate projects or reallocate resources. What seemed to be promising turned out to be a lot of busy work, with increased meetings and deliberations without clarity of outcomes. What is amiss in this situation? What can be done to make portfolio management useful?

In each of the above examples, what's missing is the right focus on strategy. In the first scenario of the energy company, effective execution was the focus up to this point; now in a disruptive world, they need to recreate their strategy to compete to win. Winning needs to be infused into execution to make it strategic-execution. In the second example of the healthcare PMO, the light bulb moment was the understanding that the PMO needs to go beyond execution and connect to the growth strategy of the organization. Things started to turn quickly as the PMO repositioned and retooled itself to focus on enabling and supporting the growth strategy. What was amiss in the portfolio management initiative of the bank was the missing link of the selected projects in the portfolio to the overall strategy of the bank. This is a common flaw where a lot of effort goes into generic portfolio selection and prioritization models without clarity on the organization's strategy.

This chapter will further delve into these examples to show how strategy is the foundational element of the DNA and how clarifying, connecting, and catalyzing a coherent strategy across the board is critical.

WHAT IS STRATEGY?

According to a survey of more than 500 senior executives around the world by Strategy&, PwC's strategy consulting business, 80 percent said that their overall strategy was not well understood, even within their own company.

Strategy is probably the most overused, misunderstood, and misinterpreted word in the business world. It is not just a project, but a "strategic project," or a "strategic initiative," or a "strategic PMO," but what does that mean? *Strategic* is often overused as a qualifier to convey that it is something big or important. Often, mission/vision statements or goals like "be the number one in our industry" are mistaken for strategy. One of the best explainers of strategy is Richard Rumelt, and in his book, *Good Strategy Bad Strategy,* he explains how a hallmark of mediocrity and bad strategy is a flurry of fluff masking an absence of substance. Most bad strategies are nothing more than an assortment of goals and desire that often contradict one another. Bad strategy is long on goals and short on policy or action. It assumes that goals are all you need for execution. It puts forward strategic objectives that are incoherent and, sometimes, totally impracticable. It uses grandiose jargon and phrases to hide these failings. Rumelt writes:

> Despite the roar of voices wanting to equate strategy with ambition, leadership, "vision," planning, or the economic logic of competition, strategy is none of these. The core of strategy work is always the same: discovering the critical factors in a situation and designing a way of coordinating and focusing actions to deal with those factors. A leader's most important responsibility is identifying the biggest challenges to forward progress and devising a coherent approach to overcoming them.

According to Rumelt, good strategy is not just "what" you are trying to do. It is also "why" and "how" you are doing it. He breaks down the kernel of strategy into three elements: a diagnosis, a guiding policy, and coherent action—a diagnosis of the current situation and a guiding policy that specifies the approach to dealing with the challenges called out in the diagnosis. He uses the analogy of a signpost, marking the direction forward but not defining the details of the trip. Coherent actions are coordinated commitments and resource allocations designed to carry out the guiding policy. A good strategy doesn't just draw on existing strength; it creates strength through the coherence of its design. Most organizations of any size don't do this. Rather, they pursue different objectives that are disconnected, and the incoherence causes further cracks in the organizational mirror.

Strategy Is Not Planning

In the project world, it is important to clarify another myth and distinguish strategy from planning. Roger Martin, the former dean of Rotman School of Management, writes in *Harvard Business Review:*

> I must have heard the words "we need to create a strategic plan" at least an order of magnitude more times than I have heard "we need to create a strategy." This is because most people see strategy as an exercise in producing a planning document. In this conception, strategy is manifested as a long list of initiatives with timeframes associated and resources assigned. Somewhat intriguingly, at least to me, the initiatives are themselves often called "strategies." That is, each different initiative is a strategy and the plan is an organized list of the strategies.

> But how does a strategic plan of this sort differ from a budget? Many people with whom I work find it hard to distinguish between the two and wonder why a company needs to have both. And I think they are right to wonder. The vast majority of strategic plans that I have seen over 30 years of working in the strategy realm are simply budgets with lots of explanatory words attached.

To make strategy more interesting—and different from a budget—we need to break free of this obsession with planning. Strategy is not planning—it is the making of an integrated set of choices that collectively position the firm in its industry so as to create sustainable advantage relative to competition and deliver superior financial returns.

How do you integrate all this and create a simplified strategy? Roger Martin along with A.G. Lafley in their book, *Playing to Win,* provide a template with five questions: (1) what is our winning aspiration; (2) where will we play; (3) how will we win; (4) what capabilities need to be in place; and (5) what management systems must be instituted?

These questions, which can be addressed on one page, provide a sound framework to practice and connect strategy to execution and other areas of the DNA from a project and PMO standpoint. The questions start with a focus on "winning," which helps to shift focus from just execution and getting things done to results and outcomes. If you are part of winning, you need to know what game you are playing and with whom, and you have a sense of ownership and accountability for results and outcomes. Your perspective shifts from execution to strategic execution with not just delivering on time and budget, but how can we win by executing with a purpose to win—how could our project delivery capabilities become a competitive advantage? Instead of a doer, you become a driver—a strategic partner and influencer on how to win by building and enabling PM and PMO capabilities. This is exactly what was missing in the above example from the energy company, where they had matured their execution over the years. Like typical companies, they had a 2020 goal of producing X gigawatts of energy, which they were well underway to achieving. They were focused on execution, and winning was not a part of their framing. They found themselves disrupted in a changing landscape of increasing competition.

Donald Sull and Kathleen Eisenhardt write in their book, *Simple Rules:*

Developing a strategy and implementing it are often viewed as two distinct activities—first you come up with the perfect plan and then you worry about how to make it happen. This approach, common though it is, creates a disconnect between what a company is trying to accomplish and what employees do on a day-to-day basis.

They provide a way to bridge this gap between strategic intent and actual implementation in three steps, which is particularly useful for projects and PMOs:

Figure out what will move the needles.
Choose a bottleneck.
Craft the rules.

The two needles (Figure 4.1) represent the top line of revenues and bottom line of costs. The first step is to identify the critical choices that will drive a wedge between revenues and costs to increase profits and sustain them over time. The second step is to identify a bottleneck, a decision, or activity that is preventing the company from improving results. The final step is to craft a set of simple rules that, when applied to the bottleneck, impacts profitability. The metaphor of the needles and the related visual of the top line and bottom line resonates well in my experience and is a very effective way

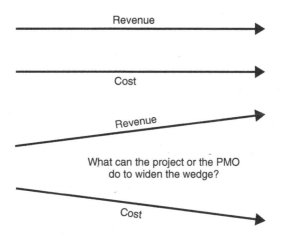

Figure 4.1: Move the Needle and Widen the Wedge

of communicating strategy and what actions need to be taken to connect execution to strategy.

The key question for any project and PMO is to ask what action we have to take to move the needles to widen the wedge. In the case of PMOs, it can be hard to show a direct impact on revenue; however, a good place to start is to see what the bottlenecks are that can lower the bottom line of cost. According to Sull and Eisenhardt, the best bottlenecks to focus your attention on, share three characteristics:

1. They have a direct and significant impact on value creation.
2. They should represent recurrent decisions (as opposed to "one-off" choices).
3. They should be obstacles that arise when opportunities exceed available resources.

Based on these bottlenecks, you collaborate with stakeholders to craft the rules that impact decision making and prioritization. Instead of a top-down governance committee crafting the rules, it is vital to include people who are involved in day-to-day activities who have project field intelligence and can provide insights that may not be obvious at higher levels.

If you are trying to win a car race, you can apply strategic thinking by trying to move the needles by either building a faster car or trying to gain efficiencies. This is exactly what Audi did when it developed the R10 TDI car for the famous 24-hour Le Mans race in 2006. Audi asked a how-to-widen-the-wedge question, "How could we win Le Mans if our car could go no faster than anyone else's?" It diagnosed the challenge and identified the bottleneck as pit stops. As Audi focused on the design elements of the problem, it led toward the answer of fuel efficiency to put diesel technology into the race cars for the first time. Audi could win the race with a car that wasn't faster than any of the other cars with the strategy of fewer pit stops. And it was right: the R10 TDI placed first at Le Mans for the next three years.

In the case of the previous energy company, it was jolted by a wave of changes in the offshore wind power industry that it had dominated for over 25 years. The company spent a lot of time diagnosing its current challenges with declining costs, new competitors, declining margins, and oversupply in the supply chain. It identified the bottleneck as the delivery mechanism, as the competition shifted to cost of electricity. The company decided that it was not going to play in the onshore market. The whole company was mobilized in clarifying that strategy to compete on cost. The project managers, led by the PMO, became key players in mobilizing this new strategy. With their technical expertise, instead of a heads-down execution approach, the PMO tapped into the field intelligence of project teams to identify the levers to influence cost and innovate on reducing the delivery cost of electricity, where the future competition was headed.

Whose Fault Is It, Strategy or Execution?

A common complaint from executives is that "execution is the problem" or "our strategy is execution." The onus is on project management and PMOs to fix execution. When things go wrong, project management gets blamed. While that might be true at times and there is always room to improve execution, often failure is wrongly attributed to execution when the real culprit is bad strategy. It is interesting to consider the following combinations:

Good Strategy + Good Execution = Win
Bad Strategy + Bad Execution = Lose
Good Strategy + Bad Execution = Possible to Win
Bad Strategy + Good Execution = Lose

If you have a bad strategy or no strategy, excellence in execution is not going to make a difference. On the other hand, companies with a good strategy can survive and navigate their way to winning in spite of weak execution. It is also important to emphasize that good strategy can create the right conditions for successful execution. The opportunity is for project management and PMO to be a player in both, strategy as well as execution and understand its role in supporting, enabling, and facilitating strategy while focusing on execution.

WHAT IS THE ROLE OF PROJECT, PROGRAM, PORTFOLIO, AND PMO IN STRATEGY?

Projects and programs are the vehicles to implement strategy. PMOs, as well as project and program managers, can play a key role in strategy. Traditionally, they have been separated as the doers versus the thinkers who focus on strategy. A central point of this book is you cannot separate the two—they are the two intertwined core elements of the DNA. For PMOs, it has been a slow evolution from being boxed into execution and unawareness of strategy to awareness, alignment, support, and participation and having a seat in some organizations at the strategy table. There is a shift from planning, administration, coordination, and support to project/program managers and PMOs being a strategic player and participating and influencing strategy. This was the case with the healthcare PMO introduced in the beginning of the chapter that had the insight to connect to the growth strategy and had started to play a key role in strategy. The PMO helped to test and validate the business hypothesis of each proposal. They challenged themselves to see how they could design and configure execution as a strategic advantage. They continued to widen the wedge by actively engaging in winning new contracts, as well as fine-tuning processes to lower costs.

Whether it already has a seat at the strategy table or not, the PMO can start to earn a spot by understanding, clarifying, supporting, aligning, facilitating, enabling, and catalyzing strategy, in its own way. You are not going to automatically get invited; you have to demonstrate that you have what it takes to play in strategy work. In a Projectize Group survey, 55 percent of PMOs reported they are involved in strategy. This is an opportunity for the PMO to not only link strategy to execution but also be a player in driving better strategy and strategic execution. Besides focusing on portfolio management, PMOs can play a role in strategic decision support by providing diagnosis, choices, and design for coherent actions based on their project and program insights and field intelligence.

THE DNA STRANDS OF STRATEGY

To reinterpret strategy with a PM and PMO lens, we have identified five strands of strategy listed in Figure 4.2.

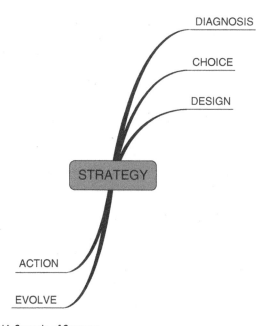

Figure 4.2: The DNA Strands of Strategy

These strands are an attempt to integrate the wisdom of the strategy gurus already discussed and apply it from a PM and PMO standpoint. Table 4.1 provides a way to apply these strands by asking the related questions and tying it to PM or PMO functions. It can also be used to develop PMO strategy intelligence. It is important to point out that there is bound to be overlap, as these strands are typically intertwined, and the order of questions or activities may not necessarily be sequential.

Strategy as Design

Strategy is often linked with decision making, but it has more to do with design. Fred Collopy and Richard Boland of the Weatherhead School of Management eloquently explain the difference between decision attitude and design attitude. The decision attitude, they write, assumes that it is easy to come up with alternatives but difficult to choose between them. The design attitude, in contrast, assumes that it is difficult to design an outstanding alternative, but once you have, the decision about which alternative to select becomes trivial. In a clearly defined and stable situation, when the feasible alternatives are well known, a decision attitude may work fine. But in a disruptive and dynamic environment, a design attitude is more effective—it can withstand the turbulence and uncertainty. Design in strategy is concerned with finding the best possible set of coherent actions given the overall challenge and the available skills, time, and resources. Project management and PMOs can apply design attitude first, by understanding and using organizational strategy to design the right criteria and boundary rules for selection and prioritization of initiatives. Second, they can translate strategy into appropriate action by designing execution elements for optimized results. (Refer to Chapter 5 for a detailed discussion on design elements in execution.)

HOW CAN THE PMO LINK STRATEGY WITH EXECUTION?

I remember my first experience as a part of a PMO almost 20 years ago. The PMO dealt mostly with delivery and execution issues in a professional services organization, in a division of about 2,000 engineers/consultants and 150 project managers. We were involved in establishing PM standards, training, mentoring, and supporting projects for two and a half years. We were busy and comfortable in our execution circle. The challenge was that the project success rate had not gone up. Some of the functional managers were complaining that the PMO was slowing things down and not adding any value. Some of them felt the PMO should be disbanded and the team should be made billable on client projects.

I clearly remember the moment of my awakening to the business and strategy sphere. As things came to a head and we were going through a reorganization, our senior vice president didn't know what to do with the PMO. He was being pressured to disband the PMO; he was not quite sure, but he saw some value and called a meeting. There was a lot of deliberation about the challenges and the role and value of the PMO from the various stakeholders. Toward the end of the meeting, the senior vice

Table 4.1: The DNA Strands of Strategy

Strategy Strand	Project/Program/PMO Questions	Activities
Diagnosis	What is the critical challenge (problem)? What are the pain points? What are the right questions? Why are we doing this? What is our winning aspiration? Where and why do we make money? What will move the needle? What future do we need to plan for? What is the diagnosis of the situation? How can we provide better diagnostic information to management and key stakeholders for effective decision making?	Portfolio analysis Project/program results, benefits, value analysis Risk and assumption assessment Identification & diagnosis of stakeholder and customer pain points & experience Assess & analyze project/program business alignment gaps Resource & capacity capability analysis
Choice	What is our guiding policy? What is our overall approach and integrated strategy? What are the potential pathways to winning? Where will we play? What will we say no to? What are the risks and assumptions? How can we help craft clear choices? How can we better understand and align project, program, and PMO to organizational strategic choices?	Facilitation & clarification of business goals & objectives Facilitation & clarification of portfolio governance & Boundaries classification, selection & prioritization & Balancing— criteria & models Execution analysis to question what will move the needle & identify opportunities?
Design	Are we asking the right questions? What are the bottlenecks? How can we address the unmet needs of our customers? How will we win? How can we better anticipate outcomes? How can we help design simple rules, guidelines, and processes for coherent action?	Frame execution with a winning perspective Seek and explore alternatives for execution optimization Effective resource allocation and capacity planning Dependency and interdependency analysis across initiatives, programs, and projects Connect and link initiatives, programs, and projects to test coherence and identify redundancies

(Continued)

Strategy Strand	Project/Program/PMO Questions	PMO Activities
Action	What capabilities must be in place? What are the steps required to carry out policy? What changes in people, power, and procedures are needed? Is our action coherent and integrated with the overall approach? How can we magnify and multiply the effectiveness of our resources and capacity?	Resource allocation and capacity commitment Identify opportunities for execution optimization Portfolio balancing Recommend actions for project termination Recommend actions for cost savings and cost avoidance
Evolve	How do we adapt and learn? Was our hypothesis valid? How do we pivot and evolve? How can we enable exploration, growth, and innovation? What are opportunities for greater customer adoption and enhanced customer experience? What are possible untapped opportunities? How can we learn and evolve and provide better decision-support?	Results, benefits, value, and impact assessment Lessons learned and retrospectives Identify opportunities for greater customer adoption and experience Support and enable exploration, innovation, growth opportunities

Table 4.1: (Continued)

president challenged us that the PMO had been in existence for two and a half years, and the value was questionable. So he challenged us to come up with a business plan for the PMO. Up to that point, I had never thought about the PMO from a business perspective. As I reflected on the challenge and worked on creating a business plan for the PMO, the light bulb went on, and I understood that there is another side besides delivery and execution of projects and PMO. It is really about the business and how the projects and PMO can impact the business and strategy.

In our approach, we recommend that every PMO should have a business plan or a business model canvas (explained below) that shows how the PMO is going to support, enable, and impact the business, with a value proposition in terms of cost and return on investment (ROI). The PMO has to show how it is going to move the needles and widen the wedge between revenue and cost. An easy way to create a business-focused project and PMO culture is to challenge and train everyone with the WIGBRFDT—What-Is-the-Good-Business-Reason-for-Doing-This? If anybody involved in projects cannot answer this in less than 30 seconds, they are wasting the company's time and resources.

To understand the business and strategy, you must start with the business model. The business model is like a blueprint for a strategy to be implemented through organizational structures, processes, and systems. It describes the rationale of how an organization creates, delivers, and captures value. Another approach that we have adopted in recent years is to create a business model canvas (BMC). Initially proposed by Alexander Osterwalder, the BMC (Figure 4.3) is a strategic management and lean start-up template for documenting and visually describing the four main areas of a business: customers, offer, infrastructure, and financial viability. The canvas illustrates the nine basic building blocks of how a company intends to make money:

1. Customer segments
2. Value propositions
3. Channels
4. Customer relationships
5. Revenue streams
6. Key resources
7. Key activities
8. Key partnerships
9. Cost structure

You can print out the BMC on a large surface and collaborate with your team to jointly start sketching and discussing business model elements with Post-it notes or board markers. It is a hands-on tool that fosters understanding, discussion, creativity, and analysis. You can download it from the Strategyzer AG Website.

Every PMO team should do this exercise of completing and validating the BMC with key stakeholders to make sure they understand the building blocks of their business and overall strategy. They should review each of the building blocks and identify opportunities to see how they can impact and add value in any of the nine areas. They can use a design approach to see what areas can be tweaked and optimized to create strategic advantage. The more the links and overlaps of the PMO activities on the BMC the greater the opportunity for the PMO. If the PMO cannot easily identify the areas of impact, it is going to be challenged.

There are many different variations of the template. We have seen our clients adapt in a variety of ways. We have also modified it to create a one-page initiative review template, in lieu of a business case for an initiative or project. Figure 4.4 is an example of an initiative review template.

Why not also adapt this for existing projects, to make sure the project team understands the business reasons for the projects and what areas of the business the project is going to have an impact. This is a practical approach to promoting business alignment and a sense of ownership in business outcomes, leading toward strategic execution.

You can take the output of the BMC and input it in a simple project or PMO business alignment matrix to show what the project or PMO is working on, and how it aligns to the business and strategy.

The business alignment matrix (Figure 4.5) is also a good way to craft your elevator pitch, as it has all the key elements that executives are interested in. What are you working on? Why are you working on it? What area of the business model does it address? Who cares, or who is going to be impacted? And how will we measure it?

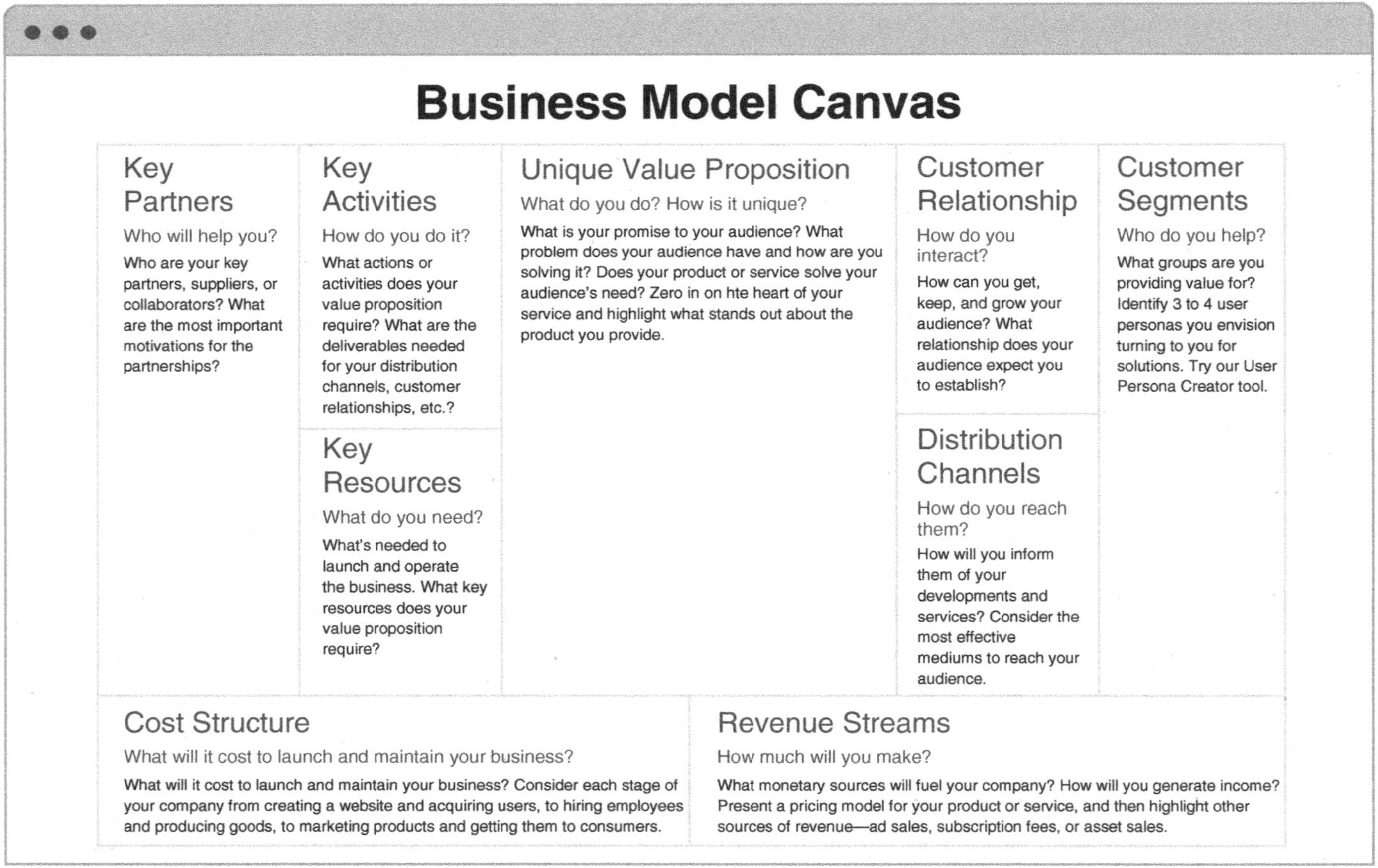

Business Model Canvas

Key Partners

Who will help you?

Who are your key partners, suppliers, or collaborators? What are the most important motivations for the partnerships?

Key Activities

How do you do it?

What actions or activities does your value proposition require? What are the deliverables needed for your distribution channels, customer relationships, etc.?

Key Resources

What do you need?

What's needed to launch and operate the business. What key resources does your value proposition require?

Unique Value Proposition

What do you do? How is it unique?

What is your promise to your audience? What problem does your audience have and how are you solving it? Does your product or service solve your audience's need? Zero in on hte heart of your service and highlight what stands out about the product you provide.

Customer Relationship

How do you interact?

How can you get, keep, and grow your audience? What relationship does your audience expect you to establish?

Distribution Channels

How do you reach them?

How will you inform them of your developments and services? Consider the most effective mediums to reach your audience.

Customer Segments

Who do you help?

What groups are you providing value for? Identify 3 to 4 user personas you envision turning to you for solutions. Try our User Persona Creator tool.

Cost Structure

What will it cost to launch and maintain your business?

What will it cost to launch and maintain your business? Consider each stage of your company from creating a website and acquiring users, to hiring employees and producing goods, to marketing products and getting them to consumers.

Revenue Streams

How much will you make?

What monetary sources will fuel your company? How will you generate income? Present a pricing model for your product or service, and then highlight other sources of revenue—ad sales, subscription fees, or asset sales.

Figure 4.3 Business Model Canvas

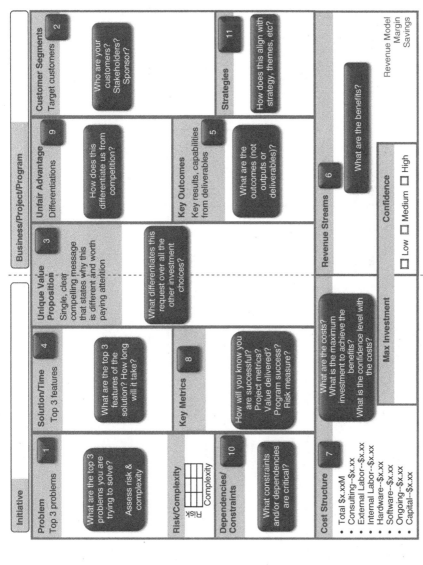

Figure 4.4   Initiative Review Template

81

BUSINESS ALIGNMENT MATRIX [WITGBRFDT?]

BUSINESS OBJECTIVE (WHY)	STRATEGY (WHY)	Business Model Canvass Areas	Project or PMO Focus Area DELIVERABLE (WHAT)	STAKEHOLDERS (WHO)						ASSUMPTIONS	METRICS (HOW WILL WE KNOW)
				Cust.	Sr. Mgt.	Fin.	Mktg.	HR	IT		
Increase revenue by 10%	Prioritize & focus on low-cost / high-margin initiatives	Cost structure Revenue streams Value proposition Key activities	Facilitate & establish portfolio mgt. process	X	X	X	X	X	X	Stakeholder buy-in / agreement on process Availability of data	No. of projects terminated due to not meeting portfolio criteria Overall cost savings
?	?		?							?	?
?	?		?							?	?
?	?		?							?	?

Figure 4.5 Business Alignment Matrix

82

SELECTING, PRIORITIZING, AND BALANCING: PORTFOLIO MANAGEMENT

If strategy is about determining where to play, how to win, and what to do and, more importantly, what not to do, then selection and prioritization of initiatives and projects in a coherent way becomes a crucial function. The accelerating DANCE and disruption have created a crisis of prioritization. To deal with the barrage of initiatives and projects, many organizations have adopted project portfolio management (PPM). Touted as a strategic function of project management and PMOs, often it is not strategic at all. It is mostly a capacity planning and budgeting exercise. Ironically, strategy and strategic dialogue is what's missing from portfolio management. Even agile prioritization is based on business value, but how do we know if it is strategic?

Remember the example of the bank that implemented portfolio management at the enterprise level. They had a good portfolio process and seemed to do everything right, according to the standard portfolio management playbook. But they struggled to make tough decisions to terminate projects or reallocate resources. They had objective ranking criteria, but there were endless deliberations. Project proponents were prone to rig the system to justify their initiative and rank it higher. As we diagnosed the problem and spoke to their executives, it became obvious to everybody that what was missing was a strategy. There was no unified understanding of strategy; they made all the classic mistakes about strategy discussed above. They confused vision and goals for strategy. The objective project ranking criteria they had was a generic model. That's why it did not make sense to the portfolio committee, and there was no real buy-in for it, even though everybody had agreed to it in the beginning.

As Lafley and Martin point out, "Strategy tells you what initiatives actually make sense and are likely to produce the result you actually want. Such a strategy makes planning easy. There are fewer fights about which initiatives should and should not make the list because the strategy enables discernment of what is critical and what is not." The original selection criteria included a common combination of (1) financial measures of ROI and net present value (NPV), (2) cost, and (3) risk.

After we worked with the executive team and applied the strands of strategy, by asking the appropriate questions, they were able to **diagnose** the pain. Now, they had a different perspective, and it became clear to them that they had to focus on cost. They outlined their **choice** to minimize spending, and focus on growth in one segment where they had an advantage. After much deliberation, they **designed** a new selection and prioritization criteria—anything that (a) removes bottlenecks in the growth segment, (b) requires minimal up-front investment, or (c) reuses existing resources. This new criterion was used after and on top of initial rankings provided by the PMO based on financial, cost, and risk analysis, to decide and commit appropriate **action.** Overtime, they **evolved** to adapt and fine-tune the criteria based on evolving strategy.

When strategy is well understood it is interesting to observe how it shifts perspective. There is more focus and fewer arguments and people understand what needs to be done to win, and why their initiative may not be most important. They know what to say yes to, and what to say no to. They can see the overall picture, to design, tweak, and commit to coherent actions.

PPM has evolved in recent years with a multitude of models, criteria, and scoring mechanisms for selection and prioritization. As you adapt portfolio management for your organization, the scoring and ranking criteria can easily multiply and get too detailed and complex, losing the essence of portfolio management. The approach that is adapted may not be relevant to your organization or too complicated. As a result, PPM is not well understood and may not have buy-in and engagement from all stakeholders.

Following are pointers on how to take PPM to the next level and infuse it with strategy:

Adapt a shark-tank approach

Traditional PPM is a demand management or capacity planning exercise with a cost and budget mindset. You need to shake things up with a shark-tank approach, to shift to an investment mindset and create a competitive and winning spirit. What if it were your own money—what kind of pitch you would want to see; how would you make investment decisions? One thing is for sure: you would focus on your investment strategy, and there would be more rigor and preparation. What is your investment strategy? What is your competition? How are you going to win?

Stop greasing the squeaky wheels

How do you make the best use of limited resources? How do you avoid the tendency to be pulled into projects that seem urgent on the surface but are not important strategically? It is important to understand the concept of triage, which is a system used by emergency personnel to sort patients according to the urgency of their need for care. Triage is designed to maximize survival and make the best use of limited medical resources. In the organizational context, triage is defined as the assignment of priority order to projects based on where funds and other resources can best be used, are most needed, or are most likely to achieve success. In organizations with limited resources and many initiatives, the "squeaky wheels" often get attention at the cost of high-value strategic projects. To avoid being pulled in the wrong direction, triage can help you establish categories based on objective criteria.

Define the right categories

To do a fair comparison and prioritization of projects, they have to be classified in appropriate categories. According to the PMI Standard for Portfolio Management, categorization is the process of organizing projects and programs into relevant business groups to which you can apply a common set of decision filters and criteria. Categorization is the precursor of the portfolio management processes of evaluation, selection, prioritization, and balancing of the portfolio.

In portfolio management, three generic categories are common at a high level: run, grow, and transform the business. Run the business focuses on the short-term, with projects related to operational continuity or the so-called "keeping the lights on" category. Grow the business is about long-term efficiency improvement projects, for example, to increase production or reduce cost. Transform the business is strategic, and focuses on projects that provide competitive advantages, such as adding new revenue sources or new positioning.

For example, in pharmaceuticals, common categories include product improvement, new platform, and breakthrough projects. In information technology (IT), the portfolio can be categorized into four investment classes: infrastructure, transactional, information-producing, and strategic class systems, according to Peter Weill, MBA, PhD, director and senior research scientist at the MIT Sloan Center for Information Research. The use of categories should help you slice your portfolio into investment classes that make sense for your organization and help you manage the portfolio effectively. The key is to make sure the classification is relevant to your business and make sure it helps highlight the important characteristics and distribution of the portfolio.

Design and adapt the right criteria

The essence of portfolio management is to help you make the right investment decisions based on evaluating risk versus reward. Project portfolio management adds resource availability and capacity as well as strategic alignment to the equation. The analysis of risk, rewards, and resources by itself is not purposeful unless the projects fit with the strategy and align with the business objectives.

Figure 4.6 illustrates the five factors that combine to form the basis for sound PPM selection criteria.

Rewards are the performance potential or any combination of benefits that may include financial criteria such as ROI, payback period, NPV, and internal rate of return (IRR). Rewards also include nonfinancial or intangible benefits, such as customer satisfaction, reputation, goodwill, or effectiveness—for example, the effectiveness of a vaccination program.

Risks represent the probability of success or failure, based on factors like length, scope, technology, complexity, and so on. Modern portfolio theory (MPT) defines an optimal portfolio as the one that generates the highest possible

Figure 4.6: PPM Selection Criteria

return for a given level of risk. MPT was introduced by Harry Markowitz in his paper, "Portfolio Selection," published in 1952 in the *Journal of Finance*. Even though MPT focused purely on financial portfolios, project portfolio management is based on similar principles. According to MPT, expected risk has two sources: investment risk, which is the risk of the individual stock, and relationship risk, which is the risk derived from how a stock relates to other stocks in a portfolio.

In project portfolio management, there is a strong emphasis on the evaluation of projects individually. Often overlooked or misunderstood is the relationship risk: you also need to **assess the impact of individual projects on the others in the portfolio**.

Resources are any combination of people, available funds, and budget, equipment, hardware, software, or any other assets required to do the project.

Strategic fit addresses how projects fit into the mix of related projects in the portfolio.

A project may appear to be optimal; it may have high benefits, low risk, and low resource requirements. However, if it does not fit into the strategic mix, it can diffuse or water down the portfolio's focus.

Example: Your IT portfolio consists of one type of technology platform, and you introduce a new project based on a different technology that is not part of the overall IT strategy. The new project has the potential to become a high-risk project and cause a ripple effect on the overall portfolio.

Strategic business objectives guide the alignment of projects with your organization's overall objectives such as cost reduction, revenue growth, increased market share, and so on. Projects should be linked to one or more of these objectives, which provide a focus to align and balance the portfolio.

Tweak and adjust the weights for each criterion

All criteria are not equally important, and you may have assigned weights, but are the assigned weights appropriate and valid? The weights determined initially might need to be recalibrated as objectives or business conditions change.

Adjust and balance the right mix of criteria

Often, there is too much emphasis on quantitative financial rewards and not enough importance given to qualitative criteria like customer satisfaction, reputation, morale, prestige, effectiveness, and so on. The goal should be to have a good mix of criteria with balance between:

- Quantitative and qualitative
- Tangible (hard) and intangible (soft)
- Direct and indirect
- Short-term and long-term perspectives

Focus on the relationship risk and impact of selecting one project over another

It is common to score the projects and rank them in a spreadsheet format. But this only provides individual project analysis, which is not enough. The next step is to plot these projects to show how they compare against each other. This is typically illustrated using multivariable bubble charts comparing two or more

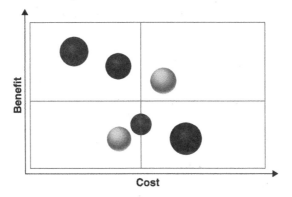

Figure 4.7: Multivariable Bubble Chart

factors. For example, Figure 4.7 depicts cost versus benefit. The bubble size is the resource requirements, and the color of the bubble represents the project classification. These charts help to visualize the relationships and impact across projects. A medical device company found it challenging to implement portfolio management in their product development PMO, as their PMO director explained, "We have four product families, and each of them has its priorities, and then we have inter-product family conflicts." Besides creating separate charts for each product family, and a consolidated visual at the enterprise level, the key was to help identify cross-project interdependencies and a way to manage them. The challenge is that some of the projects in flight can't advance far enough because of interdependencies with others (for more on managing interdependencies, see Chapter 5). These charts also help provide inputs to balance the portfolio across categories.

Develop and communicate the strategic road map

The strategic road map should reflect the strategy and objectives with a list of the prioritized strategic initiatives. This should be communicated across the organization to provide transparency and emphasize the strategic priorities. The strategic roadmap helps in fostering a holistic approach and a sense of participation and ownership in the overall organizational strategy.

Test and view PPM with strategy lens

To ensure that PPM is indeed strategic, seek to clarify and understand strategy and test and challenge strategic assumptions. Test and challenge your hypothesis to make sure you have the right diagnosis, and your criteria are based on a coherent approach. Look for opportunities to optimize, balance, and provide better strategic decision support.

Have the right PPM governance, with the right people engaged at the right level

You should develop the right governance approach collaboratively with the portfolio group, PMO, or investment committee responsible for portfolio management. Otherwise, you can expect limited adoption, undermining of the process, or gaming of the scores to squeeze through favorite projects. Part of portfolio governance is to have a stage-gate process for effective portfolio monitoring throughout the project lifecycle to facilitate go, no-go decisions, based on progress reports and health-checks. Governance mechanisms should also be used to drive objective decision making.

Learn and adapt

Remember, there is no single best approach. It is important to fine-tune and adjust as you try to find the right PPM approach that is suitable for your organization. One size misfits all, and it is better to learn and adjust, rather than stubbornly follow the process that does not work. Try different orientation and viewpoints of the portfolio mix to gain different perspectives. No matter how hard you try, there will always be a degree of gaming due to pet projects and politics. Sometimes there is a genuine reason to keep some projects under the portfolio radar due to confidentiality or security reasons. The lesson we learned was to create a separate "stealth" category—call it category X or whatever you want. This provides a vehicle for these projects to at least be identified, without all the details. At least this way they are on the radar and people know X projects are underway and that's why capacity is maxed out.

Resource Management: Is It Possible to Defy The Laws of Physics?

Have you ever seen a situation where you have more resources than projects? The reality is that people try to defy the laws of physics and squeeze more and more despite limited capacity. Insufficient resources are the biggest barrier to effective management of strategically relevant projects, and on average 29 percent of projects have too few resources given their level of importance, according to PMI. Are you spreading your resources thinly, like peanut butter across myriad initiatives? The Peanut Butter Manifesto was an internal memo written by a Yahoo SVP to describe this problem and urging the company to narrow its focus and clarify its vision as far back in 2006.

Resource challenge is a problem not just of quantity and quality, but also people and politics and lack of strategy. This reality will not change soon; it will only get worse in a more and more DANCE-world. Figure 4.8 illustrates the real-world challenge of the impact of resource requirements and capacity planning in an uncertain world.

So what can you do? Instead of fighting it and blaming resource management, learn to accept it and become a master at finding ways to optimize the resource constraint based on strategy. PMOs can add value by turning the resource constraint into a strategic opportunity. Following are next generation pointers to help deal with resource challenges for the PMO:

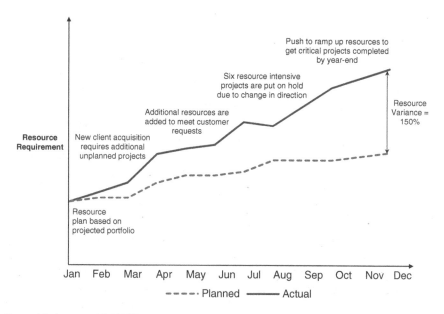

Figure 4.8: Impact of DANCE on Resource Planning

Get real

Get real about the true capacity and demand. Purely relying on tools to estimate capacity does not provide a realistic picture. Resources are people and estimating the availability and productivity of humans is different than planning capacity on a factory floor. People have different levels of productivity depending on their skills, work preference, motivations, and other factors that can be hard to nail but must be considered. A formula to consider in estimating resources is to use effort duration times availability, divided by productivity. Likewise, with demand, question, whether you have the full spectrum of demand that's competing for resources.

Turn resource constraint into opportunity

It might sound like a cliche, but this is how the PMO can add value to the resource management process. Review the overall process and diagnose inefficiencies in the system and identify any idle or excess capacity. Review existing availability and productivity issues, study the bottlenecks, see if there are ways to leverage untapped skills or availability. Find ways to better align supply with demand in the overall business model. See how you can make this painful process easy, simple, transparent, and fun. This is how companies like Uber, Airbnb, and others disrupted their industries by capitalizing on resource inefficiencies in the system. You are bound to tap into new opportunities and a whole new perspective on resource management as you shift from viewing resources as a constraint to an opportunity.

Get aligned to the right thing

Alignment is a buzzword that is overused, but the question is aligned with what? Resources are typically aligned with departments or projects and programs. Alignment by department causes siloed-thinking and a fractured approach. To break silos, alignment to projects and programs is used, but a better approach is to first align with strategy. Make sure the strategy and business model is well understood. Clarify alignment of resources to products and outcomes, instead of projects and programs activities. This is a subtle shift but can help create a sense of stake and ownership to outcomes. If the strategy is clear and everyone can see the big picture, there are bound to be fewer arguments about resource allocation.

Robust, responsive, resilient resourcing

In a dynamic and unpredictable DANCE environment, how can we plan resource management effectively? The traditional models rely on detailed up-front planning in their delivery-chain. Cemex, a global building materials company, a few years ago had an insight that led them to rethink their resource planning and cement delivery-chain, starting in Mexico City. Earlier, they had a detailed planning and scheduling of cement delivery the day before, but typical risks of traffic, weather, and so on can cause delays, and in the case of cement it can cause wastage and cost issues. As they considered how to make their supply chain more efficient, robust, and responsive, they came up with an idea to redesign their delivery model based on the emergency management system in many cities. The ambulances are stationed in strategic locations around the city and are routed in real time based on demand and traffic, weather, and other considerations. Similarly, the challenge for the PMO is to question how to rethink your resource management strategy and make it more robust, responsive, and resilient based on real-time demand and availability.

Learn to say no

Until you don't know what to say yes to and what to say no to, and how to say it, resource management will remain a challenge. The next section illuminates the problem of project termination and the importance of saying no.

What Will You Say No To?

"The essence of strategy is choosing what not to do," according to the strategy guru Michael Porter of Harvard Business School and echoed by many others. This is also the hardest to practice, to have the discipline to choose what not to do. Project termination is an area of weakness commonly cited by PMO respondents. In a study by PMI, respondents believed that 20 percent of current projects in strategic portfolios should be terminated, yet limp on, using valuable resources.

It is tough to say no or kill projects for a multitude of reasons, including sunk cost fallacy (costs already incurred and cannot be recovered) and being at a point of no return, mixed with politics, pet projects, and legal ramifications. This does not just happen to company projects, but also high-visibility projects involving governments like the Concorde project. The British and French governments continued to fund the joint development of the Concorde even after it became apparent that there was no longer an economic case for the aircraft. The project was regarded privately by the British government as a "commercial disaster" that should never have been started and was almost canceled, but political and legal issues had ultimately made it impossible for either government to pull out. This phenomenon has been named the "Concorde Fallacy" by evolutionary biologists as a metaphor for when animals or humans protect an investment—a nest or project or program—even when keeping it costs more than abandonment.

A study on the strategic role of portfolio management by Sergey Filipov et al. sums up what is at stake:

> Portfolio management is successful when "you can kill a project" (explicitly meant, using precise and objective criteria and processes). Strange as it may seem, the answer unveils the essence of portfolio management - to ensure that an organization is doing the right projects right, and the wrong projects wrongly executed can be effectively eliminated.

The PMO can play a key role in enabling the decision making by implementing objective criteria and processes for portfolio management, and more importantly providing and communicating actionable information. One way to practice the art of saying no and make people think about the consequences of adding more and more is to ask, 'the strategic question' coined by Michael Bungay Stanier, "What will you say *No* to if you're truly saying Yes to this?"

PMO's Role in Portfolio Management

According to a Projectize Group survey of over 500 PMOs, 48 percent of PMOs are responsible for the Portfolio Management function in their organization. Regardless of whether your PMO is directly responsible or not, following is a checklist of how the PMO can facilitate and support PPM:

- Define and develop the detailed, continuous process by which projects are evaluated, prioritized, selected, and managed.
- Support and facilitate portfolio governance.
- Foster a collaborative effort that enables senior executives and the portfolio committee to reach agreement on portfolio objectives.
- Provide coaching and training to project managers to help them to understand project evaluation criteria and to enable them to efficiently generate inputs for the project template.
- Communicate to project proponents which projects are approved and project priorities.
- Negotiate and coordinate resource conflicts between projects and programs.
- Provide different views and analytics to enable effective portfolio decisions.
- Seek to make PPM strategic and ensure that the project portfolio is aligned with strategy and business objectives.
- Identify lessons learned and continually refine the portfolio management process.

ENABLING EXPLORATION AND INNOVATION

How can the PMO facilitate the balancing of the portfolio? Traditionally, there have been a number of models and matrices like the Boston Consulting Group (BCG) matrix, which assessed the value of the investments in a company's portfolio since the 1970s. The four-field matrix categorized investments as cash cows (milk as long as you can), stars (invest), question marks (tough decision), and dogs (terminate). Other models that have been used more from a project portfolio investment standpoint have included white elephants—which take a lot of care and feeding but won't succeed or have much benefit; bread and butter—keep the lights on projects that pay the bills; oysters are the incubators—we don't know if they will succeed, but they are promising; and pearls are the gems that we want to invest in. These types of matrices have been commonly used to balance the portfolio with risk versus reward, comparing factors like probability of success versus potential benefit.

The challenge is how to classify and manage in today's DANCE and disruptive world.

As you are pushed to explore and innovate and grow rapidly, resources are spread thinly, and there is a crisis of prioritization. Start-ups tend to be focused from this standpoint, whereas existing companies try to do too much and lack focus. Geoffrey Moore, in his book *Zone to Win: Organizing to Compete in an Age of Disruption,* offers a framework for managing disruptive innovation. This framework is helpful for PMOs to make sure they are focused on the right things to enable and facilitate exploration and innovation and have the right execution approaches for each zone.

According to Moore, this framework is based on two principles:

1. Disruptive innovations (which involve new technologies, new markets, and new business models) need to be managed in a categorically different manner than sustaining innovations (which, however "new" they are, do not fundamentally change the business).
2. Innovations that have a material impact on current revenue need to be managed in a categorically different manner than back-office services and experimental initiatives.

Applying these two principles organizationally results in four zones of activity:

1. **Performance zone.** Current-year revenue performance (PMO opportunity: fine-tuning execution performance)
2. **Productivity zone.** Productivity initiatives to foster and fuel performance (PMO opportunity: optimize, compliance, efficiency, effectiveness)
3. **Incubation zone.** Incubation of future innovations (PMO opportunity: positioning to enable and support the next big thing)
4. **Transformation zone.** Taking innovations to scale (PMO opportunity: enable to scale the new disruptive initiative to scale, making sure it drives focus on the one big transformative initiative)

The classic misconception about innovation is that it is limited to only product innovation. However, as Larry Keeley et al. point out in their book, *Ten Types of Innovation*, there are multiple ways to innovate, besides focusing on the product. They list 10 types of innovation—profit model, network, structure, process, product performance, product system, service, channel, brand, and customer engagement. This is particularly relevant to project management and PMO, where they can play a role in innovation and add value. They can look for innovation opportunities in process, service, and customer engagement innovation, and to a degree in impacting profit model with cost savings or revenue impact. In the next chapter, we will see how project management and the PMO can link and leverage strategy by building an adaptive execution platform to enable and support innovation in process, service, customer engagement, profit model, and other potential innovation areas.

Innovation often is a result of serendipity and happens by accident and can't be planned for. Project management and PMOs are about execution, which requires planning and focus but also can prevent new discoveries. So how do you cultivate a serendipity engine to foster innovation? Joe Ito, director of the MIT Media Lab, talks about developing peripheral vision, which can prepare you to spot opportunities that you might miss if you are just focused on execution. The trick is to develop the ability to be able to go back and forth between the two. Besides supporting processes for performance and productivity, the PMO has to also have incubation and transformation in its peripheral vision, and make sure that the processes are not inhibiting but enabling innovation. The PMO has to support what Professor Charles O'Reilly of Stanford calls "ambidexterity"—the ability of an organization to compete in mature markets and technologies, where key success factors are exploitation of execution capabilities and efficiencies and, simultaneously, exploration in emerging markets, which requires flexibility, initiative, risk-taking, and experimentation.

EVALUATING RESULTS, VALUE, AND IMPACT

> However beautiful the strategy, you should occasionally look at the results.
> —Sir Winston Churchill

How will you know if the strategy is on track? Another strategy function of the PMO is to track and evaluate, benefits, results, value, and impact. According to the Association for Project Management (APM):

> If value is to be created and sustained, benefits need to be actively managed through the whole investment lifecycle. From describing and selecting the investment, through program scoping and design, delivery of the program to create the capability and execution of the business changes required to utilize that capability, and the operation and eventual retirement of the resulting assets. Unfortunately, this is rarely the case.

It is not enough to evaluate the potential benefits of the project in the business case in the beginning; you have to measure and track the benefits throughout the life cycle. The PMO can chart the benefits realization management (BRM) process,

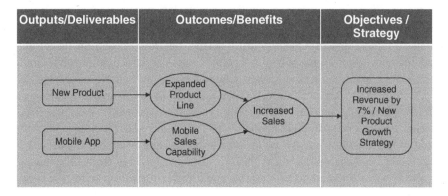

Figure 4.9: Benefits Map

provide benefit mapping templates, and educate project teams. The PMO has to create a culture where project and program managers take ownership of the outcomes and focus on both delivery outputs and benefit outcomes. We will take an in-depth look at measurement in Chapter 8. Figure 4.9 is an example, illustrating the linkage of deliverables to benefits and objectives and strategy. For the benefits to be realized, a behavioral change has to occur. In the following example, the benefit of increased sales from the two new products will be realized only if customers act and buy the new product.

Some benefits can also be perceived by some as dis-benefits or negative by one or more stakeholders. For example, the benefit of an expanded product line can increase operational costs and be a dis-benefit for operations. If dis-benefits are identified proactively, they can be managed better and even turned into opportunities, like an increased budget for operations in the above example.

The overall goal for strategic execution for the PMO is to look for opportunities to prioritize and maximize strategic benefits throughout the lifecycle. Benefits realization management should be part of the stage-gate review governance process.

DEVELOPING PROJECT MANAGEMENT AND PMO STRATEGY INTELLIGENCE

Use the following checklist of questions to assess and develop project management and PMO strategy intelligence:

- Is the definition of strategy understood and applied in the right way?
- Is the strategy clear and understood by everyone in the organization?
- How can we do better at communicating and cascading strategy throughout the organization?
- Do our selection and prioritization criteria support our strategy?

- Are the initiatives and projects selected and prioritized to align and support the strategic goals?
- How can we better understand and tap into our strategic advantage?
- How can our decisions optimize the effectiveness of our resources and actions?
- How can we create a supportive environment for portfolio management?
- Do we have the right level of sponsorship and support for portfolio management?
- Does leadership respect and value portfolio management?
- How can we do better at resource allocation and capacity planning?
- How can we get better at project termination?
- What will we say No to if we are truly saying Yes to this?
- How can we communicate effectively for tough decision making?
- How can we craft and establish simple rules for effective decision making?
- How can we improve our capabilities with appropriate portfolio management competencies, processes, and tools?
- How can we improve the accessibility, availability, and quality of data and systems for effective portfolio management?
- What effective processes and tools can we utilize to better link strategy with execution?
- Do we have effective measures and data analytics for timely project/program decision making?
- How can we reduce and better deal with portfolio politics, gaming, and pet projects?
- Do we have effective governance with built-in transparency for objective decision making?
- Does our culture support portfolio management by rewarding appropriate behaviors and calling-out weaknesses or problem areas?
- How can we provide coaching and training to project managers to help them to understand project evaluation criteria?
- How can we identify lessons learned and continually refine the portfolio management process?
- How can we better evaluate the results, value, and impact of our strategy?
- Do we have a process to manage benefits realization?
- Are we too focused on execution and exploitation, and not enabling exploration and innovation?
- How can we better enable exploration and innovation?
- How can we clarify, connect, and catalyze a coherent strategy across the board in everything we do?

KEY TAKEAWAYS

- *Strategy* is the most misunderstood and misinterpreted word. It is not vision, mission, goals or objectives. Strategy is about diagnosing the critical factors in a situation and designing a way of coordinating and focusing actions to deal with those factors.

- Projects and programs are the vehicles to implement strategy. The opportunity is for project management and PMO to be a player in both strategy and execution, and understand its role in supporting, enabling, and facilitating strategy while focusing on execution.
- Strategy is not planning—it is different than budgeting. It is the making of an integrated set of choices that collectively position the firm in its industry to create sustainable advantage relative to competition and deliver superior financial returns.
- A straightforward way to apply strategy is to question what will move the needle and what we can do widen the wedge between the top line and bottom line.
- To reinterpret and apply strategy from a project management and PMO standpoint, focus on the five strands of strategy—Diagnosis, Choice, Design, Action, and Evolve—with the application of related PMO activities of portfolio management; resource and capacity planning; results, benefits, value, and impact assessment; and support and enable exploration, innovation, and growth opportunities.
- Project portfolio management is a crucial selection and prioritization function, but often it is a capacity planning and resource allocation exercise and is not strategic. You have to raise the bar and infuse strategy into PPM by reviewing your portfolio management process and instead of using standard portfolio financial and risk criteria, identify opportunities for different categorization and criteria that impact strategy.
- Strategy is at least as much about what an organization does not do as it is about what it does. The essence of strategy is choosing what not to do. Ask the strategic question, "What are you saying no to, if you are saying yes to this?" Provide objective data and convincing analytics for justifying project termination.
- Look for opportunities to see how project management and the PMO can link and leverage strategy, by enabling and supporting different types of innovation like process, service, customer engagement, profit model, and other potential innovation areas.
- If value is to be created and sustained, benefits need to be actively managed through the whole investment lifecycle. The overall goal for strategic execution for the PMO is to look for opportunities to prioritize and maximize strategic benefits throughout the life cycle.

5

EXECUTION

Without strategy, execution is aimless. Without execution, strategy is useless.
Morris Chang, CEO of Taiwan Semiconductor Manufacturing Co.

Leading Questions

- What are the DNA strands of execution?
- What is strategic execution? How is it different than execution?
- How to assess and improve execution capabilities?
- How to plan for execution agility and scalability?
- How to develop execution intelligence?

Chances are that you have probably consumed a food item today that was transported in a transport refrigeration unit by Thermo King. It is the world leader in transport temperature control systems for trucks, trailers, buses, rail cars, and seagoing containers. For nearly 80 years, Thermo King (TK), a brand of Ingersoll Rand, has been developing customer-focused innovations for a variety of transport applications.

Their strategy is to prioritize projects and initiatives that drive market share growth, based on enhanced quality and customer focus. They have utilized project management and PMO practices in the background for a number of years. With new PMO leadership they questioned, how can we evolve project management and refocus our execution capabilities toward strategic execution and business needs?

WHAT IS EXECUTION?

If strategy is about what to do, or what not to do, then execution is about the how to do it. Simply, execution is about getting things done, from vision to reality, or from a goal to a check-mark. Execution also happens to be the top challenge in organizations. Whereas strategy has been studied, researched, analyzed, and taught in business schools and business literature, execution has not received the same level of treatment. In the last 20 years or so there has been a growing focus on project management as one of the disciplines of execution and a way to get things done, especially strategic initiatives and related projects.

Organizational project management (OPM) provides the foundation of processes and tools focused on accountability for time and cost. The focus has been on efficiency—to shorten the distance between goal and accomplishment, by focusing on how we do it faster, better, and cheaper. Just in the past few years with program management, there is a shift toward effectiveness with an emphasis on what we do that has greater benefits and value. In today's world, there is a need to go beyond and focus on experience—customer, end user, and stakeholder experience and impact.

The question is how next-generation project management and PMOs can cover and connect all these aspects in a holistic way and develop strategic-execution capabilities.

Execution versus Strategic Execution

Listing and highlighting the distinction between execution and strategic execution provides a new framework for next-generation project management and PMO. As strategy and execution come together, and strategic execution gets encoded into the genetic code of the organization, eventually the distinction should dissolve. To start with, it is important to understand the difference and identify opportunities to evolve (Table 5.1).

Out of the many PMOs, we have worked with, there are a few who understood this. While many PMOs focus on standardization and consistent delivery, the Thermo King PMO has simplified its mandate as "executing to deliver business results." Over the

Table 5.1: The Difference between Execution and Strategic Execution	
Execution	**Strategic Execution**
Getting things done (task-focus)	Getting things done with an outcome focus (executing with the big picture in mind)
Siloed execution	Integrative and holistic execution platform
Making sure we are doing it right	Questioning whether we are making the right choices and doing the right things
Mechanical and linear execution approach	Organic and adaptive execution approach
Risk avoidance and mitigation	Risk acceptance, exploitation, and enhancement
Execution with immediate goals in mind	Execution with long-term gaze and strategic intent
Focus on performance (optimizing delivery and outputs)	Focus on performance with purpose (optimizing strategic outcomes and impact)
Execution spirit, ownership, and accountability for delivery	Entrepreneurial execution and winning spirit, ownership, and accountability for business results

past year and a half, the PMO and project management team has strived to inject an entrepreneurial spirit as they have understood and embraced the company's growth strategy and found ways to execute on it. They are prioritizing quality and voice of customer initiatives, along with new product development to deliver on the growth strategy. It is rare to hear quality, delivery, cost on one hand, and operating income, margins, and sales volume on the other, in a room full of project managers. It wasn't easy, but project managers have a different perspective now, with ownership and accountability for business results. Instead of jumping ahead into the technical design solution, they are bound to question and engage in the commercial hypothesis first. They discuss operating income for the project and how are we trending toward business results. They are also geared for adaptive execution, as business conditions change, the PMO is better tuned-in to anticipate, adapt, and adjust spending and resource allocation.

Overall project management is gaining a reputation for strategic execution, and the PMO is perceived as the key to link execution issues across silos and resolve them with a holistic perspective.

So how does next-generation project management and PMO develop strategic-execution capabilities? You need to build an adaptive platform to enable and optimize strategic execution.

DECODING THE DNA STRANDS OF EXECUTION

For effective execution think of designing and building an adaptive platform composed of four strands and connecting strands of execution listed in Figure 5.1.

PEOPLE: DEVELOPING STRATEGY EXECUTION CAPABILITIES WITH THE RIGHT TALENT

Execution is powered by people. You cannot execute without the right talent. The right combination of skills and competencies needed to get things done in a DANCE and disruptive world. In a hyperconnected and hypercompetitive environment with high-stakes initiatives, there is a greater premium on those who can execute, and the need for effective people strategy should be a top priority for effective strategy execution. Talent management should be aligned with overall business strategy. Although most executives recognize the importance of considering talent management as connected to strategy, it receives too little attention. According to a Project Management Institute (PMI) research study, only 23 percent of respondents believe that senior leadership gives strategic talent management for project and program management the priority it deserves.

With increasing complexity and pace of change, it is not enough to hire and develop technical capabilities alone. Project managers need skills beyond the traditional "triple constraint" of bringing projects in on time, in scope, and on budget.

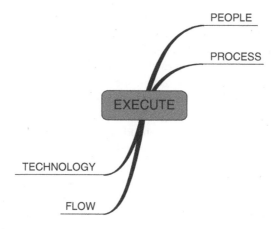

Figure 5.1: DNA Strands of Execution

There is a need to lead and direct projects and programs—not just manage them. To address this need, PMI developed the Talent Triangle, which includes strategic and business management and leadership skills, besides technical management skills, as illustrated in Figure 5.2.

Figure 5.3 lists the various skills in each of the areas of the talent triangle.

The Talent Triangle provides a broader framework for skills development. The technical skills cover some of the execution aspects, and strategy and business handle strategy from the DNA of strategy execution. For next-generation project management and PMO we need to go deeper and surface some of the subtleties and nuances that are overlooked by asking underlying questions:

How do we deal with the changing nature of the workforce and work dynamics in today's digital and hyperconnected world?

How do we develop talent that recognizes the DANCE and can learn to DANCE and thrive in the midst of it?

Figure 5.2: PMI Talent Triangle

Source: © 2010–2015 Project Management Institute. All rights reserved.

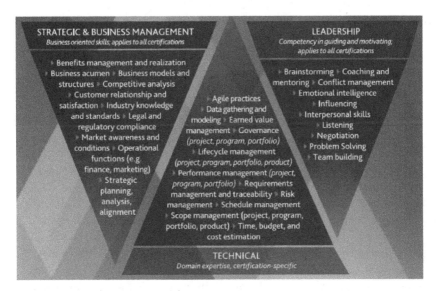

Figure 5.3: Talent Triangle Skills

How can we ensure that all elements of the DNA of strategy execution are covered in our talent development strategy?

How do we cultivate an entrepreneurial spirit and sense of ownership and commitment at every level?

To address these questions, you need to cast a wider and deeper net and aim for a holistic talent development perspective. We recommend six additional aspects in our talent assessment and development practice: (1) artistry, (2) DANCEing, (3) change making, (4) connecting, (5) learning, and (6) entrepreneurial spirit to develop next-generation project management and PMO capabilities. Think of these aspects as characteristics, like the polish that is required to transform the triangle into a multidimensional diamond over time.

Developing Next Generation Talent and Skills

Artistry

Typically, project managers do not think of themselves as artists. They have a mechanical mindset and rely on standardized processes, methods, checklists, tools, and templates akin to running the project on autopilot. They see little room for artistic expression. This is a common misconception!

Projects by definition are unique, and viewing and treating them with an artistic perspective can offer new insights to better manage them.

Rob Austin from Harvard Business School and theater director Lee Devin, authors of *Artful Making: What Managers Need to Know about How Artists Work*, note that "as business becomes more dependent on knowledge to create value, work becomes more like art. In the future, managers who understand how artists work will have an advantage over those who don't."

The mechanical approach is based on a machine-oriented view of organizations and projects, as if projects are manufactured using structured processes in a controlled environment that can deliver predictable and consistent outputs each time. The challenge is that work has changed and is changing even more—from factory to knowledge and now connected and creative work. While knowledge workers focused on either technical track or management careers, today's workers are what Eric Schmidt and Jonathan Rosenberg of Google call the "smart creative."

In their book, *How Google Works*, they describe smart creatives. They are multidimensional, usually combining technical depth with business savvy, creative flair, and a hands-on approach to getting things done. They are not confined to specific tasks. They are not limited in their access to the company's information and computing power. They are not averse to taking risks, nor are they punished or held back in any way when those risky initiatives fail. They are not hemmed in by role definitions or organizational structures; in fact, they are encouraged to exercise their own ideas. They don't keep quiet when they disagree with something. They get bored easily and shift jobs a lot.

Another point to bear in mind is the changing nature of the workforce. According to a whitepaper by Changepoint, a provider of project portfolio management (PPM) solutions, by 2020, Millennials will make up half the global labor force. By 2030, they'll account for 75 percent. They work differently and have different expectations. Millennials' aversion to hidden agendas, rigid corporate structures, and information silos coupled with a willingness to explore new opportunities will either fundamentally change the nature of work or severely cost businesses.

Next-generation project management and PMOs can be better prepared for the changing workforce by questioning how they can deliberately cultivate and tap into the smart, creative spark of their people. Instead of stifling artistic freedom, they encourage it. They hire, develop, and cross-train talent for multidimensional skills, for both mechanics and artistry. To assess the artistry capabilities of project managers and PMOs, question if they:

- Emphasize design before jumping into planning. Do they design the project with a creative eye, with a built-in flexibility to rearrange project planning based on emerging stakeholder needs?
- Are constantly curious, ask "what if," challenge the rules, and substitute or combine elements for creative solutions.
- Have the ability to empathize and focus on stakeholder/customer needs and experience.
- Take a holistic and integrative approach.
- View constraints as creative opportunities for innovation.
- Are better prepared to deal with the changes and variability of project environments.
- Improvise based on experience, insight, intuition, and judgment and make appropriate adjustments.

Art can be viewed in a negative connotation because it cannot be codified or easily understood. But that is precisely why it can offer a competitive advantage because it cannot be easily replicated. Artistry skills can become invaluable. As Seth Godin, in his book *Linchpin: Are You Indispensable?* puts it, artists become indispensable as they are the linchpins whom everybody counts on to get things done. Developing artistry can enrich the project manager's palette with skills that complement the traditional tools of project management. To be successful, you need to complement and balance the mechanical competencies with artistic talent.

DANCEing

To develop an adaptive mindset, you have to recognize and understand the DANCE. You must learn to DANCE and thrive on the edge of chaotic volatility, ambiguity, and unpredictability. Recognize the difference and adjust the approach depending on which game you are playing, whether you are dealing with a linear "pool" type of project, or you are playing "pinball" in a DANCE environment. It should be emphasized that although adaptive planning and agile techniques are useful, this is different. It is more about cultivating the right mindset and getting the operating system ready before adopting agile methods and apps. It is about having an adaptive and emergent bottom-up leadership to accomplish things without authority.

Assess adaptive skills of your talent by questioning:

- What is the degree of awareness and sensing capabilities to recognize the DANCE [Dynamic | Ambiguous | Nonlinear | Complex | Emergent] characteristics?
- Is there a recognition of which game they are playing and the appropriate approach?
- Are they more black and white or comfortable with gray?
- How comfortable are they with uncertainty and ambiguity?
- Do they recognize complex adaptive systems and their characteristics?
- Do they understand the distinction between simple, complicated, complex, and chaos?
- Do they know when to use which approach, methodology, and tool?
- Do they distinguish between process and behavioral complexity?
- How aware are they of human behavior dynamics?
- What is the degree of their adaptive ease-capability to sense, respond, adapt, and adjust (SRAA)?
- What is their readiness to dance on the edge of chaos—to thrive in the midst of ambiguity and uncertainty?

Changemaking

Change is the catalytic element in the DNA of strategy execution; it can either stimulate and sustain or slow and stall the project or PMO. You can excel in all the other six elements of the DNA, which provide the structure, process, and capabilities for strategy execution, but without change, you are jeopardizing adoption and intended outcomes of your initiative. That's why developing change intelligence, and related skills is imperative.

The prevalent perspective of change particularly in project management relates to managing, controlling, and monitoring change. There is a lack of ownership and responsibility for the success of the change. There is a fundamental shift that needs to happen from managing and leading change toward the idea of owning and making change happen. A changemaker isn't someone who simply manages change and wishes for change; he or she makes change happen. Changemakers can transform; they influence the outcomes through responsibility, ownership, and determination. With appropriate change skills you can evolve from a change manager and game player with technical skills, to a change leader and game influencer with strategic business skills, to a game changer with transformational changemaking skills.

Chapter 9 explains the difference between managing change and changemaking and details how to develop change intelligence and related skills with change awareness, anticipation, absorption, and adoption skills.

Connecting

Connection is the foundational networking and circulation skill that breathes life and adds complexity. Just like you cannot connect and communicate without a good Wi-Fi connection, similarly, you cannot get things done without connection. Multiplicity of connections can add exponential complexity. Project managers and PMOs have to develop skills to identify the disconnects and opportunities to bridge and strengthen the connecting nodes. Also, connecting skills are vital to decipher the connections, and their interplay that causes complexity. Chapter 7 delves into how to connect stakeholders, silos, business, and interfaces and interdependencies by developing networks and connections, marketing and communication, relationships, and community and collaboration skills.

Learning

To continuously evolve and innovate it is critical to cultivate insatiable curiosity and learning agility. It is important to not just execute but to execute and learn from it, and question how to make it better and easier over time. Chapter 10 addresses the 7C's of developing learning intelligence: culture, curiosity, capture, curation, community and collaboration, correcting, and continuous innovation.

Entrepreneurial

Another dimension that can either catalyze or hamper effective strategy execution is the degree of entrepreneurial spirit—a sense of purpose, ownership, and commitment at the individual level that is more prevalent in start-ups, to develop this aspect question: *how can we make our people think like owners?* Entrepreneurial spirit is naturally more prevalent in start-ups, and incumbent companies like CITI, Pfizer, Unilever, WestJet, and others have tried to infuse it into their talent development strategies.

Next-generation project management and PMOs need to assess and evaluate the above skills as they fine-tune their hiring, retention, and talent development strategies. Of course, there are few candidates who will have all the right mix of the above competencies.

As you build talent, make sure your team is balanced with each person with one or more of the above skills. The idea is to know the strengths and weaknesses and staff appropriately. For example, high-impact strategic initiatives that are complex might need more of the artistry and adaptive DANCEing skills.

Designing Career and Learning and Development Programs

Effective PMOs partner with human resources (HR) to develop project management roles, career path, certification, and training requirements. As you evaluate these requirements, the goal should be to create the right mindset, along with the appropriate skillset and toolset. To cater to a changing workforce, learning and development needs to utilize a variety of blended options catering to not only different learning styles but also mixed generations of digital natives. To cater to shorter attention spans, affinity for task switching and preferences for learning what is minimally necessary, a blend of formal and informal learning with micro-learning opportunities, gamification, and hands-on, live learning work better. Also, it is critical to provide live learning opportunities with built-in reflection time and deep, insightful learning opportunities. In today's world, this is rare and hard to gain from other mediums and can be a refreshing and mindful experience to generate breakthrough thinking and actionable insights.

Besides the talent triangle training on technical, strategic & business management and leadership skills, it is also important to provide learning programs in managing the DANCE and complexity, along with change intelligence, connecting, and learning agility.

Learning and development can be supplemented by mentoring and coaching programs. In my observation, PMOs do not effectively develop or utilize coaching programs for project management, like other executive coaching programs.

Also, part of people development is to have a defined career path. Figure 5.4 is an example of a project management career path from Nationwide Insurance Company PMO.

PROCESS

An adaptive platform of enabling processes can help people perform better and is foundational for execution. Identifying the right processes that are critical to execute in your business is key.

On their journey to evolve project management and PMO processes, Thermo King conducted an assessment in the following PM processes:

Stakeholder engagement and sponsorship
PM methodology standards
PPM tool management

Sample PM Career Path

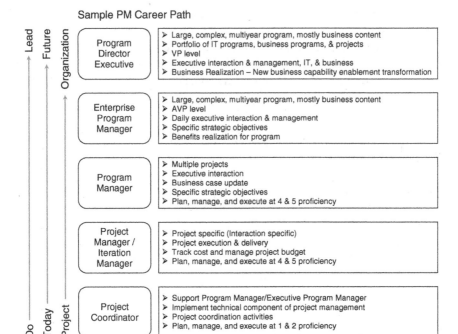

Figure 5.4: Sample PM Career Path
Source: Adapted from Nationwide Insurance Co. PMO.

Project and program delivery
Estimation
Risk management
Cost management
Resource management
Contractor management
Metrics and measurement
Portfolio management
PM talent development
PMO performance management

Initially, when they conducted the assessment, they were ranked in the middle, with a rating of 3 on a scale of 1 to 5. They used this to determine their strengths and weaknesses and come up with a PMO roadmap to fine-tune related processes.

There are many of assessment frameworks and maturity models espoused by consulting companies. For an initial straightforward self-assessment, PMI's *Project Management Body of Knowledge (PMBoK) Guide's* process groups and knowledge areas provide a foundational framework that can be used as a matrix to list and assess project-related processes.

The matrix in Figure 5.5 can be used in a number of ways: You can do an assessment in each of the process groups, by overall knowledge area, or you can list relevant

	Initiation	Planning	Execution	Monitoring & Control	Closeout	Total Score
Integration Management						
Scope Management						
Time Management						
Cost Management						
Quality Management						
Human Resources Management						
Communications Management						
Risk Management						
Procurement Management						
Stakeholder Management						
Total Score						

Figure 5.5: PM Knowledge Areas Assessment

processes in each knowledge area in a more detailed evaluation. You can assess on a scale of 1 to 5, or you can interview an individual or a group of stakeholders. The ranking can be simply 1=low and 5=high, or you can also base it on the Strategy Execution Continuum: 1=adhoc (firefighting), 2=standardize, 3=measure, 4=learn and 5=innovate, as explained in Chapter 3. The scores can help identify strengths and weaknesses and prioritize pain points.

Additionally, processes related to product development, systems development, and other cycles depending on your methodology approach, whether it is waterfall, agile, lean, or hybrid, need to be identified and developed based on your business, function, and industry.

Tips for Effective Process

It is important to emphasize that it is easy to get carried away with process, and PMOs typically get caught-up in too much process. Following are tips to consider for effective methods and process implementation:

- Select and identify only critical elements. Apply Pareto's 80/20 principle to identify the vital few processes that can have the maximum impact on execution, instead of getting overwhelmed by too many processes.

■ Focus on core principles for standards, methods, and processes, and provide room for situational application. Each organization and situation is unique, the more specific and detailed the method it may become the stumbling block, particularly for linear, logical project managers. Instead, provide principles that show them how to cook and execute, rather than detailed recipes that may be hard to follow.

■ Strive to find the sweet-spot of rigor without rigidity (discussed in Chapter 2).

■ Experiment, learn, fine-tune, and simplify processes continually.

Process versus Behavior

In our practice, it is common to come across clients that are eager to show their list of processes that they have developed, and yet we find they struggle and wonder why they are not getting the results. You can have the best documented process in the world, but if people don't understand it, or change their behavior to adopt and apply the process, it is of no use. Think from both perspectives—from the organizational inside-out perspective and the process user, or customer outside-in standpoint. Plan for three levels of process maturity and adoption. Just because you have done a great job in creating and documenting the process, it is not enough. You can't check-it off yet; it is only the first step. Table 5.2 lists the three levels of depth from both perspectives.

Classifying Processes Provides New Insights for Effective Execution & Process Innovation

Classifying processes helps to provide insights for the best usage and advantages and limitations of each category of process. Sigurd Rinde, founder of Thingamy AG and blogger, has an interesting way to classify process as bespoke, barely repeatable process, easily repeatable process, and automated process.

Bespoke processes are custom processes that are unique to your business and need to be custom-tailored each time. These handcrafted processes aim to provide a unique and memorable experience for customers.

A nonlinear unpredictable barely repeatable process (BRP) uses common ways of doing things, but still requires a high level of skill and judgment to execute. Services,

Table 5.2: Three Levels of Process: Organizational/PMO versus User/Customer Perspective		
Level	Organizational/PMO Perspective	User/Customer Perspective
Level I	Documented—written and documented	Awareness—aware that there is a process and how to access it
Level II	Communicated—communicated, trained, and supported	Understood—understand it and know how to use it
Level III	Implemented—implementation steps and support and related governance and measurement	Adopted and applied—practice, apply, and integrate the process in their day-to-day and find it useful

support, health, government, education, and all those other "people" processes that require human judgment and situational application are examples of BRP. These are supported by to-do lists, meetings, collaboration tools, and management involvement.

A linear predictable easily repeatable process (ERP) is one that doesn't require much skill or advanced training to complete; you can be trained in how to run that process with instructions. Production, procurement, distribution, and operational processes are examples of ERP that are supported with systems.

Automated process is when a human doesn't need to interact with a system to produce the outcomes. If you have a standard procurement process to order supplies each month that is an ERP that can be automated easily by setting a recurring ordering transaction.

According to Sigard's analysis, only 30 percent of the world gross domestic product (GDP) is created with ERP, whereas 64 percent is generated by BRP. While it is relatively easy to gain efficiencies from ERPs, BRPs, are different and therein lies the opportunity for exponential gains by focusing on them and finding ways to optimize these processes. This classification can help in a number of ways: to gain further efficiencies in ERP and find ways to further automate them. BRPs can be identified as processes that need greater support, training, and attention to identify optimization opportunities and augmentation with analytics and AI.

Over time, the experience from bespoke processes can be used to standardize and create signature processes that can be scaled, like Apple's product development process or Amazon's synchronized supply-chain process. These signature processes can be a competitive advantage, as they are hard to emulate. This can be an opportunity for incumbent companies to capitalize on process innovation and tap into signature processes that others like start-up companies may not yet have.

Also, the process categorization can help to evaluate appropriate systems, tools, and governance approaches for different processes.

PMO Support and Services Related to Execution

Effective PMOs provide execution support and services like Southern California Edison (SCE) Utility companies' Smart Connect program. The SCE PMO provided services like risk management workshops and organizational readiness reviews. The risk management workshop process included multiple workshops focused on potential system and product failures. A cross-section of department and directors and managers were represented. These workshops resulted in response and recovery plans for the highest probable risks. The PMO organizational readiness team works closely with operational and business area subject matter experts to assess the impact of change associated with new technology and processes, and identify key readiness activities to ensure a smooth transition. This team provides support to impacted business area leads through process walk-through events to ensure the enterprise-wide business units understand their roles and responsibilities and are prepared to assume ownership of new business process implementation.

PMOs can provide the following support services to facilitate and enable effective execution:

- Project start-up workshops
- Planning workshops
- Risk management workshops
- Organizational readiness review
- Project review and guidance
- Lessons learned and retrospective facilitation
- Troubled project recovery

TECHNOLOGY: TOOLS, SYSTEMS, APPS, AND BOTS

Technology is the enabling element of the execution platform. Having a defined technology platform and tools strategy is necessary for effective execution.

The right combination of tools and systems can enable a robust execution platform. We can easily get enamored with glitzy charts and dashboards and the appealing benefits of tools. While tools are meant to accelerate momentum, they tend to take over and impede execution if implemented incorrectly. The risk is that you can get caught up in managing the tools instead of managing the business and projects.

One of the large banks we worked with did a six-month due diligence and evaluation process to select an enterprise PPM tool. For over eight months the PMO struggled to implement various modules, with limited adoption. After spending over $1.8 million, they had to give it up and look for other options. In our experience, this scenario is common. The question is, why does this happen?

To understand the challenges of tool implementation, it is important to highlight common tool-related myths:

1. The tool will solve all the problems. (Tools cannot fix underlying systemic issues and pain points.)
2. The tool is the PMO. (The tool is only a supporting element. Projects or PMOs cannot be run on autopilot with a tool, at least not yet.)
3. We can tailor the tool to our needs. (It is not that easy to customize complex workflows unique to your organization.)
4. If we just implement it, it will work. (It takes a lot of groundwork, preparation, and time to make it work.)
5. We can get all the information that is not available today. (The data has to be input into the system. Only the data that is in the system can be accessed—it will not magically appear in the tool if there is no way to collect it.)
6. Everybody will adopt and use the tool. (Adoption rate for these tools is hard, and it is not guaranteed that people will use it.)

Address the following questions as you plan your technology platform and tools strategy:

- Do we have a technology platform roadmap and tools strategy?
- What kind of toolset is appropriate for our organization?

- Is our culture ready for the tool?
- Is it the right timing to implement the tool?
- Do we have processes and means to collect data for the tool?
- How will this tool integrate with other tools and processes?
- How can we prepare ourselves for the tool implementation?
- What will we get that we cannot get today without the tool?
- How is the tool going to impact strategy execution?

Types of Project Management and PMO-Related Tools

Project management software covers a range of solutions that allow individuals and teams to track the progress of projects from their conception to their completion. There are hundreds of these programs on the market, ranging from basic free online programs all the way to highly complex products that allow the user to manage every aspect from idea generation all the way to invoicing, payment, and benefit tracking.

According to the softwareadvice.com website, PPM refers to a collective effort by an organization to centralize and coordinate project efforts across a portfolio of work. PPM software helps automate processes, streamlining the planning, managing, and delivery of each project. While traditional project tools are designed to support teams at a project or department level, PPM software is designed to support project processes at an organizational level. For example, traditional planning allows managers to map out a project's critical path and identify task dependencies and constraints. PPM software does this and more, allowing decision makers to map out the dependencies and constraints between projects and identify potential scheduling issues, budget conflicts, and overlapping objectives. Ultimately, this allows us to better weigh one initiative against another so the business can eliminate waste and execute on the projects that deliver the greatest value. Implementing a full-fledged PPM suite requires a significant investment of time and resources to implement and manage.

The right timing to implement a PPM tool is when you have standardized and achieved a degree of consistency and repeatability of project-related processes. People understand the foundational concepts and baseline data is being collected. The culture and PM community should be ready and hungry for the tool. As management guru Jim Collins has remarked, technology should be used as an accelerator of momentum and not creator of momentum.

Tools Evaluation Resources

Several organizations specialize in providing reports and reviews for enterprise software tools. Refer to the following resources to research and evaluate tools options:

Gartner (gartner.com). Gartner is an IT research firm that publishes research reports on a variety of software categories including project management. Research reports and consulting are available for purchase.

Forrester (forrester.com). Similar to Gartner, Forrester is an IT research and advisory firm that publishes research reports and guides on enterprise software. Reports and specialized advice are available for a fee.

Softwareadvice.com. Free website that offers a wealth of evaluation information and buyers' guides on all kind of software tools including project management and PPM. They also publish a Frontrunners Quadrant, a graphic of top-performing tools in each category.

Capterra.com. A crowd-sourced software review website covering many different software categories.

You may not be ready for full-fledged PPM tools. At the same time, you don't have to rely on old-school static spreadsheets or PowerPoint. There are a lot of cool visual, mobile, and intuitive and useful apps for task management, workflow, and collaboration that are cloud-based and relatively inexpensive to tryout. Research the above websites like softwareadvice.com for the latest technology landscape.

Next-Generation Tools with Project Intelligence (PI)

The next generation of evolving technology and tools platform have sensors and bots that collect and provide live data. With the proliferation of the Internet of Things (IoT), sensors and bots capture data in the areas of quality, effort, and performance, We can imagine how timesheets and data entry is becoming antiquated with bots and assistants that help with estimates, budget, and sprint management among other things. As you work with different systems, it can track what systems you are working on for how much time. With machine learning, it can track your number of interactions as well as the content of interactions. It automates the assignment of resources and tasks. Your personal bot mines tasks from meeting notes and mark completion in real time, while the estimation bot provides realistic estimates based on mining real-time data of like projects in similar industries.

This also addresses a big challenge, as we don't have to rely on humans anymore to enter accurate and timely data. Sensors are collecting data automatically. Systems are getting smart by learning from task description and filling in the blanks by making good enough assumptions, for example, tying together sprint history, and showing how your key resource is being pulled away each week for other projects. A system can reassign tasks, based on its knowledge of how good people are based on their performance in real time. Overtime PI will move from descriptive and diagnostic domain of project management to providing predictive, prescriptive, and intuitive insights. PI will not only diagnose and predict project issues, but also resolve them intuitively before they happen, and impact project performance in real-time.

FLOW

For things to get done, work has to flow through organizational channels. Part of designing the execution platform is to map the workflows and identify the inputs, outputs, interfaces, and touchpoints across people, process, and tools that can either enhance or impede execution. Sigund Rinde articulates a useful analogy of how workflow is no different from a flow of water, that is, if you want to make good use of the water, or workers, it requires a framework to steer the flow in the right direction. And workflow frameworks, like water flow, can be categorized in three ways—pipeline, riverbed, or bucket passing.

Linear, predictable flows or ERPs, can do well with pipelines, like assembly lines or case handling, enhanced with processes, governance, and automation tools. But for nonlinear, unpredictable flows or BRPs, which inherently also have DANCE characteristics, pipelines are inadequate. What's needed is a riverbed type of an organic framework, where the riverbanks keep the flow going in the right direction. Water is fluid and adaptive and finds its way around each rock or hurdle. The challenge has been that with a mechanical mindset, it is hard to think organically and design adaptive alternatives. Also, we did not have the right technology, so organizations have resorted to structures and processes akin to bucket passing, which is inefficient, slow, and leaks time and value. Today, we have workflow and collaboration tools that can do better with BRPs. But next-generation project management and PMOs can add value by designing an execution platform by focusing on flow—the interdependencies and interfaces that can accelerate or impede flow.

Interdependencies are the established or logical reciprocal relationships between people or things that determines the behavior or performance of the whole. Interfaces are the functional and physical characteristics required to exist at a common boundary or connection between persons, between systems, or between persons and systems. Area, surface, or function provide and regulate contact between two elements of a system, or a common physical or functional boundary between different organizations or contractor's products. It is usually defined by an interface specification and managed by a system integration organization. Interface management is the management of communication, coordination, and responsibility across a common boundary between two organizations, phases, or physical entities which are interdependent according to the PM glossary (maxwideman.com).

The broader the identification of interdependencies and interfaces within project, across projects, programs, and portfolios and overall organizationally, as well as externally with vendors, partners, and customers, the more robust the execution platform. These interdependencies and interfaces can be classified as pooled, sequential, or reciprocal workflows. Richard Daft, in his book, *Organizational Theory and Design*, defines them as follows:

Pooled workflows. Accomplish tasks independently of others. All tasks must be completed to achieve desired outcomes and no timing or technical dependencies between tasks. Everyone can work regardless of what else is going on or whatever anyone else is doing and can be done anywhere.

Sequential workflows. Start or complete tasks when others have completed prerequisite tasks on which they depend. The doers have a dependency to understand their position in the sequence and what occurs after them. Also, there is the added task of defining the sequence and communicating it. Nothing can be done unless and until the doers get the communication and have the opportunity to ask clarifying questions to generate understanding adding to the complexity of defining, communicating, and managing the required sequence. While baseball can be characterized as an example of pooled interdependence, football is sequential.

Reciprocal workflows. Tasks can only be accomplished through collaboration/ negotiation with others via a series of iterations involving mutual adjustment. For example, you run into a problem with a resource; they do not have the right level of skills needed to get an important task completed. You contact Julie, the resource manager, and she informs you that she has to check availability and can get back to you by tomorrow. You have to wait for a day and find out that she does not have additional resources. You check if you can use vendor resources, and she says you have to contact Margaret for approval and submit a resource request form. Note the collaborative aspects and the involvement of multiple people, processes, and systems. Reciprocal ones are the ones that have BRP characteristics and are hard to manage. In the future, tools with AI and machine learning will do better at adaptive flow, but identifying critical interdependencies and interfaces is foundational.

Use the template in Figure 5.6 to identify and track interdependencies and interfaces.

For each of the interfaces and interdependencies, design appropriate people, process or tool resolution strategy that can enhance the flow. Overall, the PMO can promote a culture of interface and interdependency awareness and focus on flow, and look for opportunities to remove any blockages that impact flow and impede execution.

Project	Interface / Interdepen-dency	Type (Pooled / Sequential/ Reciprocal)	Input Needed	Who Needs the Input?	Who Supplies the Input?	Output	Owner/ Date Needed	Flow Resolution Strategy
XXX	Go/no go decision	Pooled/ sequential	Input data	Sponsor Gov. Committee	PM Team members	Report/ updated dashboard	Sponsor/ (xx-xx-xxxx)	Timely data input Automated collection Intuitive tool Training - tool/process
YYY	Resource requirement	Sequential/ reciprocal	Resource request & justifica-tion	Resource manager	Project manager project team	Resource	PM/ (xx-xx-xxxx)	Defined process Collabora-tion platform Resource capacity transparency

Figure 5.6: Managing Interface and Interdependency Matrix

Flow can either block or lubricate execution. Well-designed flow, powered by technology, can eliminate unnecessary documentation and approvals, based on built-in rules and machine intelligence and reduce cycle time and drive execution.

DESIGNING AN ADAPTIVE EXECUTION PLATFORM

For an adaptive execution platform that can withstand the DANCE, you need to think design. How do you optimally design the execution elements of people, process, technology, and flow? In project management, there is an emphasis on planning and not as much on design. Rushing into planning without design is like detailing the engineering blueprints of a building without thinking about architecture. While the plan spells out the details of the project, the design provides the form, function, and structure to organize it.

Planning in a DANCE environment is like dancing on a constantly moving and changing landscape with a great deal of uncertainty and unpredictability. A sound design is adaptive and can better withstand the dynamic changes and uncertainty. Design utilizes an architecture approach—providing form, function, and structure to ensure the feasibility and viability to enable the vision of the project or program. Design focuses on the structural interfaces, linkages, and dependencies. Design is based on a holistic integrative approach, as opposed to reductionist and breakdown basis of planning.

It is important to distinguish planning and design. Table 5.3 lists the difference between a planning versus a design mindset.

As you assess and develop each of the execution elements, think about how you can start with the design elements listed in Table 5.3. The application of design thinking has flourished in many industries. Tim Brown, CEO and president of IDEO, an

Table 5.3: Planning versus Design	
Planning	**Design**
Engineering approach—spells the details and provides a mechanism to execute the vision	Architecture approach—provides form, function, and structure to ensure the feasibility and viability to enable the vision
Focus on tasks and activities	Focus on interfaces—linkages and dependencies
Focus on what needs to be done	Focus on why it needs to be done
Geared toward deliverables and outputs	Geared toward experience, optimization, and outcomes
Hierarchical organization	Visual and contextual organization
Reductionist breakdown approach	Holistic integrative approach
Analytical process	Creative process

innovation and design firm, describes design thinking as "a discipline that uses the designer's sensibility and methods to match people's needs with what is technologically feasible and what a viable business strategy can convert into customer value and market opportunity."

A design mindset will help you to optimize the interplay of people, process, technology, and flow to build an adaptive execution platform. Apply the principles of design thinking to cultivate a design mindset.

Principles of Design Thinking

Following is a summary of some of the design thinking principles that can be applied to project management and PMOs culled from the literature based on the work of Tim Brown and David Kelly of IDEO; Roger Martin, dean of Rotman School of Management at the University of Toronto; Jeanne Liedka, Darden School of Management at the University of Virginia; and Tim Ogilivie of Peer Insight:

Feasibility, viability, and desirability. Match people's needs with what is technologically feasible and what a viable business strategy can convert into customer value. Intersection of feasibility (what is functionally possible within a given time frame); viability (is it viable from a business standpoint) and desirability (this is what customers, end users, and other stakeholders want).

Aim to connect deeply with those you serve. This involves observation, gaining insights and deep careful understanding of what customers want, walking in their shoes, validating assumptions, asking penetrating questions, deep listening, connecting emotionally, and having empathy for your customers and end users. This is different than conducting a business requirements process or interviewing stakeholders or conducting focus groups. It is insightful observation, living in their world and experiencing their perspective to design and deliver what they want. Ironically, projects rely on classic scope definition and requirement-gathering processes that are often static and cursory, relying heavily on documentation. Design thinking espouses that it is hard for people to tell you what they want; you must observe, engage, and empathize with them to gain insights.

Constraints as opportunities, not limitations. Design thinking by nature is divergent, open to possibilities, and seeks opportunities. It advocates not letting your imagined constraints limit your possibilities. An oft-used example is the constraint of buildings getting proportionally heavier, weaker, and more expensive as they grow larger in size. Buckminster Fuller was inspired by this constraint to invent the geodesic dome, which becomes proportionally lighter, stronger, and less expensive as it grows larger in scale.

The traditional project management mindset is the management and limitation posed by constraints of time, cost, and scope. Instead of limiting what we cannot do, design thinking help us reframe the problem and discover new opportunities in the process. According to Richard Buchanan, former dean of Carnegie Mellon School of Design, great design occurs at the intersection of constraint, contingency, and possibility.

Visualization, prototyping, and iteration. Design thinking visualizes opportunities and believes that you can't achieve perfection in a rapidly changing world and emphasizes quick prototyping and frequent iteration. It utilizes visual tools and relies on testing, seeking feedback and refining.

The above principles are well suited to address the execution challenges in a DANCE environment and essential to design an adaptive execution platform.

Project and program managers can also use this type of thinking to design projects for maximum benefit and intended outcomes within the given constraints and boundaries. It provides a greater opportunity to understand and focus on what customers and key stakeholders need and design a plan to deliver it. For example, in a large IT systems implementation project, the rollout was typically planned on a regional basis. After a design approach was introduced, the project team realized that a better plan was to implement systems based on a line of business or departmental basis. In the previous approach, if there were implementation problems, the whole site would be down. In the new customer-focus designed plan, the site would be operational and business could be conducted even if there were issues in one of the areas.

Designing for Execution Agility

Detailed execution plans considered solid today may not survive the DANCE of future uncertainty. The customer thinks that the project manager has everything under control and the project will go according to plan. The project manager is confident of the plan, but the rigid plan may blind her or him to recognize a shift in the customer's needs and the need to adjust the plan.

To deal with this, the ideas of "rolling-wave planning"—detailed planning for the immediate wave and high-level milestone planning for future waves—became popular. Also, practices like iterative planning, progressive elaboration, adaptive planning, and agile techniques are used more and more. Although these techniques can help, a fundamental shift from a "solid" to a "liquid" frame can set the stage for a more effective collaboration between a customer and a project manager.

This is important because it sets the stakeholder's expectation for variation and helps project managers by preparing them to withstand unforeseen changes and adjusting plans accordingly instead of proceeding with a false sense of comfort.

Following are ideas on how to prepare for agility and cultivate an agile mindset:

Think liquid, not solid

Plans are based on the understanding of current assumptions and objectives, which are bound to change. Rarely do projects go exactly according to plan. Working without a plan is not wise, but following a plan that has no relationship to reality is foolish. Recognize how solid plans can become blinders that limit your options. Learn to develop liquid plans with multiple perspectives and options that are enabling and not constricting. A pilot flying a plane with a clear objective starts with a flight plan, but must constantly replan in real-time based on weather conditions and wind patterns to reach his destination. Similarly, successful project managers start with a plan but manage and adjust the project in real-time based on unfolding reality on the ground.

Sense-respond-adapt-adjust

Real-time planning can be enhanced by cultivating skills to sense, respond, adjust, and adapt. Sensing skills help you develop acute awareness and vigilance to anticipate unexpected changes and respond accordingly. Adaptation helps you to quickly adjust to new realities and alter the plan to accommodate the changes.

Use the appropriate communication approach to set the right expectations

Stop using the word *plan*; substitute it with *planning*. If they ask for the plan, tell them here is the latest "planning" as of now. This subtle shift infuses a sense of fluidity, elasticity, flexibility and agility. Do not talk about planning without listing "assumptions" or "what must be true" for this planning to work.

Designing a Scalable Adaptive Execution Platform

An adaptive execution platform must be scalable. You should think about scalability and self-replication as you design your execution platform. Question how scalable are your people, process, technology, and flow. Salim Ismail, Michael Malone, and Yuri Geest, in their book, *Exponential Organizations,* use the acronym SCALE to outline a framework for rapid scalability:

Staff on demand. Use contractors, staff-on-demand platforms wherever possible; keep full-time equivalents to a minimum.

Community. Leverage PM and PMO communities to scale. Get access to additional skills and resources. Get feedback and learn to scale.

Algorithms. Identify processes and data streams that can be automated. Investigate what insights and algorithms can be learned from existing data and systems to scale.

Leveraged assets. Do *not* acquire assets unless you have to. Use cloud computing and cloud-based PPM tools and apps that can be scaled easily.

Engagement. Design PM processes and tools with engagement in mind. Gather all user interactions. Gamify where possible. Create a digital reputational system of users and suppliers to build trust and community. Use incentive prizes to engage crowd and create buzz.

These tips are more oriented for start-ups, but similar ideas, including concepts based on the lean start-up movement are being adopted within large incumbent organizations like GE and Coca-Cola. As you design for scalability, think about how incumbent companies can have an advantage from integrating an established set of capabilities in powerful ways that start-ups may not be easily able to emulate.

Assessing and Maturing Execution Agility

An adaptive execution platform with standardized processes, tools, and flows can provide stability in a DANCE-world. When everyone speaks the same language, understands how things are done around here, knows who does what, with ownership and accountability and the right level of governance, it creates a vibrant environment for agility. If you have to reinvent the wheel each time, agility suffers. Chapter 3, introduced a Strategy Execution Continuum for assessing maturity and developing intelligence:

Standardize for developing capabilities and achieving consistency and repeatability.
Measure to get visibility and transparency to drive performance and results.
Learn to improve predictability, while developing experience and agility to evolve.
Innovate for better customer experience and customer creation and sustained value
  generation and impact.

BUILDING EXECUTION INTELLIGENCE

Use the following checklist of questions to assess and develop project management and PMO execution capabilities and intelligence:

- How can we better articulate and ensure that the distinction between execution and strategic execution is understood and ingrained?
- How can we assess, develop, and balance the talent triangle—technical, strategic, and business management and leadership skills?
- How can we deal with the changing nature of the workforce and work dynamics in today's digital and hyperconnected world?
- How can we develop talent that recognizes the DANCE and can learn to DANCE and thrive in the midst of it?
- How can we cultivate an entrepreneurial spirit and sense of ownership and commitment at every level?
- How can we assess, develop, and balance artistry skills?
- How can we emphasize and further develop change-making, connecting, and learning skills?
- How can we fine-tune hiring, career path, and retention strategies to reflect strategic execution?
- How can we redesign learning and development (training) to develop next-generation capabilities?
- How can we fine-tune processes and methods to focus more on desired principles and less on how-to?
- How can we go beyond documenting processes to better implement and mature our processes from both an inside-out organizational perspective, and an outside-in user perspective?
- How can we use process categorization to evaluate appropriate systems, tools, and governance approaches for different processes?
- How can we optimize signature processes and integrate our execution capabilities for unique advantage?
- How can we better utilize technology and tools?
- Do we have a technology platform roadmap and tools strategy?

- How can we improve and redesign flow to remove blockages for effective execution?
- How can we design our execution platform for scalability?
- How can we apply the Strategy Execution Continuum to assess overall maturity in execution?

KEY TAKEAWAYS

- If strategy is about what to do, or what not to do, then execution is about the how to do it.
- Listing and highlighting the distinction between execution and strategic execution is insightful and provides a new framework for next-generation project management and PMO.
- For effective execution, design and build an adaptive platform that is managed by people and powered by technology and processes, with unimpeded flow.
- Besides technical, strategic, and business management and leadership skills, you must also assess, develop, and balance artistry, DANCEing, changemaking, connecting, learning, and entrepreneurial skills.
- You can have the best documented process in the world, but if people don't understand it, or change their behavior to adopt and apply the process, it is of no use.
- Develop and optimize signature processes and see how you can integrate execution capabilities for unique advantage.
- Categorize processes as bespoke, easily repeatable (ERP), and barely repeatable (BRP) processes to evaluate appropriate systems, tools, and governance approaches for different processes.
- Flow can either block or lubricate execution. Well-designed flow, powered by technology, can eliminate unnecessary documentation and approvals, based on built-in rules and machine intelligence and reduce cycle time and drive execution.
- An execution platform with standardized processes, tools, and flows can provide stability in a DANCE-world. You can mature it and make it intelligent by focusing on measure, learn, and innovate.

6

GOVERNANCE

Robust governance should make it possible to travel the highways of strategy execution in clearly marked lanes at high speeds while minimizing risk and enhancing performance and agility.

Leading Questions

- What is governance?
- What should be the purpose of governance?
- What are the DNA strands of governance?
- How do we implement adaptive governance?
- How do we develop governance intelligence?

Over one-third of projects failed to achieve their goals. Each week, executives had approximately 30 hours of project meetings to attend. Even worse, each project manager ran the meeting differently and provided different reports. Decision making was fragmented with too many people involved in multiple levels of approvals. There was no clear process for executives to know what was expected of them. It drained a great deal of time from executives for little result. In a word, the approach was chaotic. The project management office (PMO) along with the project leaders were heads-down trying to get things done, buried in resolving the day-to-day issues, disconnected from business strategy. All this resulted in low sales numbers on a new product, and increasing customer complaints and product recalls in this global industrial products company. It didn't take much for us to recognize the pain in this scenario was emanating from a lack of good governance.

Imagine if you had to drive like the old days on dirt roads that are not paved, on a terrain that is cracked and made up of different silos, each with its own rules. There are some signs, but they are confusing and hard to decipher. People drive on different sides of the road as they follow their own standard without a common code of communication. There are no traffic lights or markers and no warning signs to prevent risks. Everybody is going in different directions, and without any alignment, it is cumbersome and slow. There are frequent clashes along the way, as people finger-point and argue. You have been told to drive faster and keep track of time, and make sure you stay within the budgeted fuel limit. But you have no way of measuring or managing your performance. While you are busy trying to execute and get there, you have a limited idea as to why you are driving to begin with.

If this seems too far-fetched, think again. This is akin to what happens in many organizations due to lack of good governance. The different vehicles are like the projects and strategic initiatives in organizations. These projects are meant to be the strategy-execution vehicles to take the organization from the as-is to the to-be, future state. But without good governance, it is hard. There is no alignment between strategy and execution; there are disconnects due to lack of connection and communication; there is weak measurement and hard-to-monitor performance; and there is lack of preparedness for risk, change, or the ability to learn and evolve.

As the above scenario illustrates, ineffective governance risks poor quality and higher costs. According to the Project Management Institute (PMI) report, *The High Cost of Low Performance* (2016), "$122 million [is] wasted for every $1 billion invested due to poor project performance, a 12 percent increase over last year." Poor governance contributes to poor performance and ineffective decision making. But the higher price is business losses, reputation damage, or weakened competitive position and failure to realize benefits to drive innovation.

TO DEFINE EFFECTIVE GOVERNANCE, UNDERSTAND ITS PURPOSE

The word *governance* is pervasive, but it is often misconstrued and not well understood. Governance is not management or execution. Management is the decisions you make while you are executing; governance is the structure for making them. While you are driving, you are making decisions; the lanes, the markers, and the signs provide the governance—the structure and boundaries—so you can steer in the right direction, without getting in trouble.

Corporate governance is the set of processes, customs, policies, laws, and institutions affecting the way a corporation (or company) is directed, administered, or controlled. Project management or PMO governance is the subset of corporate governance focused on project, program, and portfolio management.

PMI's Practice Guide, *Governance of Portfolios, Programs, and Projects* defines organizational governance as a structured way to provide control, direction, and coordination through people, policies, and processes to meet organizational strategic and operational goals. And organizational project management (OPM) governance is defined as "the framework, functions, and processes that guide organizational project management activities to align portfolio, program, and project management practices to meet organizational strategic and operational goals."

A simple way to think about governance is as the structure for decision making, by defining and implementing standards, policies, procedures, rules, and guidelines.

Defining governance becomes effective and practical by understanding its purpose. Often, the purpose of governance is misconstrued as oversight, monitoring, and control, which prompts a framework of restrictive rules and policies. Instead, the question should be how to galvanize and invigorate strategy execution by designing a structure in which we can ensure alignment, make the right decisions, plan for risk, measure performance, and execute quickly and efficiently without getting in trouble. At the highest level, robust governance should make it possible to travel the highways of strategy execution in clearly marked lanes at high speeds with no fear of collision.

Discuss the purpose and focus of governance in your organization and the PMO by asking, how can we use governance to ensure:

Alignment with business objectives and strategy
Focus on the right priorities
Proactive risk management
Not getting in trouble
Standard methods and practices are working and effective
Benefits and business value
Responsible use of resources
Effective performance management
Effective communication
Compliance with regulations
Ethical conduct
Steering in the right direction
Adapting in the face of rapid change

THE STRANDS OF GOVERNANCE

Figure 6.1 lists the DNA strands of governance. A sound governance structure rests on the clear definition and understanding of these mutually supporting strands. Woven together, these strands lead to effective governance. You will notice that words like oversight, monitoring, and control are missing from the governance strands, as they prompt a negative and constricting view of governance. If you apply and practice these mutually supporting strands, they will result in effective oversight and control without being called out. These governance strands are generic and can be applied at the organizational project management, project, program, portfolio, or PMO level.

Steering Body

To define and establish governance, you need governing bodies like a steering committee or a board, or council, represented by various members based on areas

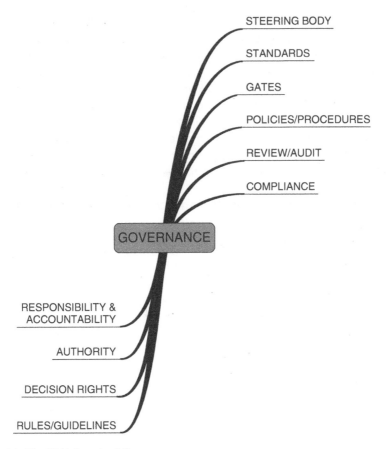

Figure 6.1: The DNA Strands of Governance

of responsibility and authority. Governing bodies are organized to formulate and establish governance criteria, and provide guidance and decision making for various aspects of acceptable governance. They take part in framing the issues and setting the boundaries. Steering bodies that are potent and impactful have a clear charter and clear delineation of responsibility without overlapping roles. It is also important to rotate committee membership and have a balanced representation of a cross-section of stakeholders. To maintain agility, the committees should be dynamic in their meeting and decision making, to expedite the flow of strategy execution.

Standards

Standardization is the first step to establish a foundation of stability in a chaotic environment. Review which industry standards might be relevant to your industry and business and how adopting them might increase your effectiveness. Standards for project, program, and portfolio management from PMI can be complemented with industry standards and methodologies. The PMO can facilitate the adoption of methodology lifecycles, whether it is traditional, agile, or hybrid. Review each of the DNA elements and see if any of the standards might be relevant. For example, in strategy, review to see if it makes sense to adapt the portfolio management standard for alignment. In execution, assess to see what aspects of the *Project Management Body of Knowledge* (PMBoK) guide can be adapted as internal standards for estimation, documentation, reporting, project management qualifications, and career path. If you are in product development or manufacturing, see if there are any applicable quality standards that might be relevant to your industry.

Policies/Procedures

For the key organizational activities in each of the DNA areas that have a direct impact on projects and programs, check to see if there are existing policies and procedures and if they are effective. Review how to define and establish or fine-tune an existing process for any of myriad activities like issues escalation, time and expenses, utilization, billing, and others, just in execution alone. Similarly, go down the list in the DNA areas and identify the high-impact processes for which effective governance can expedite the flow of execution.

Gates

Phase gates or stage gates are decision points as a project or program progresses through lifecycle phases. At each gate, continuation is decided by the governing body, like a steering committee or the sponsor. Decisions are made based on pre-established criteria like business case alignment, risk analysis, resource availability, benefits viability, and other criteria. Having a clear criterion is critical to make objective decisions. Primarily, three aspects should be reviewed at each gate:

1. **Execution**—if it meets performance and quality criteria up to this point,
2. **Business rationale**—if the project or program still holds up to the promised business rationale to justify continuation.
3. **Next phase action.** Viability of proposed action plan and resources requested for next phase.

The outcome of a gate review is a variation of one of three decisions:

1. Go (start something new—phase or project, etc., or continue to next phase).
2. No-go or kill (stop or kill because it does not meet the gate criteria).
3. Hold (hold until more information is available, or certain criteria are met).

Other variations like recycle and conditional-go can also be added. Project/ Program managers and PMOs have to prepare and be ready with the deliverables for the gate criteria. PMOs can facilitate the gates process and support the teams with action plans resulting from gate reviews. Well-defined gates help to prune the portfolio of weak projects and aid effective decision making.

Figure 6.2 is an example of what and how to decide at each gate based on implementation at the organization described in the opening scenario of this chapter.

Decision	What It Means	When to Use
Cancel (No Go / Kill) Gates 0-4	- Project MUST stop. - Archive project documents. - Team released to work on other activities.	- The strategy no longer makes sense. - The market changes. - The potential reward is not enough. - There are better places to invest resources. - The risk level is too high. - The technology is not feasible.
Hold Gates 0-4	- Project activities MUST stop until specific documented conditions are met. - Archive project documentation. - Team released to work on other activities. - Once the specified conditions are met, team will recycle the last gate.	- The strategy still makes sense, but - The funding is not available. - The resources are not currently available. - The technology is not ready.
Recycle Gates 0-5	- May allow the team to continue select activities while working to shore up or finalize other activities. The decision should be very specific on what can & cannot continue. - With a decision of RECYCLE, the gate must be repeated. - The team MUST supply a date for the repeated gate.	- The strategy still makes sense AND - The resources are available BUT - The gate deliverables have not be satisfactorily met.
Continue (Go) Gates 0-4	- Resources and spending are approved. - Project team allowed to move to the next phase of the project. - The team MUST supply a date for the next gate.	- The deliverables have been satisfactorily met. - The potential reward is acceptable. - The technology is feasible. - The risk is palatable. - There is no better place to invest resources. - There are no remaining questions regarding gate deliverables.
Archive Project and Release Team Gate 5 only	- Project has been satisfactorily completed and is ready to be passed to the sustaining team for life cycle management and metrics tracking. - Archive project documentation. - Team members released to work on other activities.	- The deliverables have been satisfactorily met. - There are no remaining questions regarding gate 5 deliverables.

Figure 6.2: Gate Decisions

Review/Audit

Project and program reviews are an essential part of any lifecycle to assess the status and see if it is trending in the right direction, or if a course correction is required. Reviews should be conducted during the lifecycle as a part of the gate review or outside of the gate review, and also at the close of the project. Depending on the industry and the nature of the project or program, it may also have to go through an audit process. PMOs can help prepare for audits. Agile approaches have built-in sprint reviews and retrospectives. Agile techniques for reviews and retrospectives (discussed in Chapter 10) are engaging and infuse a spirit of collaboration and learning to this otherwise dreaded governance aspect.

Compliance

Depending on your industry and the type of project or program it may involve legal, regulatory, financial, industry, environmental, and other compliance criteria. For example, compliance criteria related to Health Insurance Portability and Accountability Act (HIPAA), if you are in healthcare, or Sarbanes-Oxley, in banking and finance and other international compliance regulations related to security, privacy, or the environment. Project/program managers and PMOs have to be cognizant of the compliance issues and check if appropriate governance related to compliance is included and planned for.

Responsibility and Accountability

Unclear responsibility and accountability lead to confusion and delays. A commonly used technique is a responsibility assignment matrix (RAM), also known as RACI matrix, which describes the various roles for completing any activity. RACI is an acronym for who is responsible, accountable, consulted, and informed for a particular task or deliverable.

Responsible is the "doer" who does the work to complete the task. There is at least one with an "R" designation to do the work, and there can be multiple people who are "R" to get the task done.

Accountable is the "owner" of the task, the one ultimately answerable for the correct and thorough completion of the deliverable or task, and the one who delegates the work to those *responsible*. In other words, an "A" must signoff (approve) work that *responsible* provides. There **must** be only one "A specified for each task or deliverable.

Consulted are those who need to be consulted or their opinion is sought either before or in the process of accomplishing the activity—typically, subject-matter experts—and with whom there is a two-way communication.

Informed are those who are informed, often only on completion of the task or deliverable, and with whom there is just one-way communication.

RACI is a simple framework to assign roles and responsibility for each of the governance areas. Even if you don't establish a full-fledged governance framework, just establishing RACI for critical activities is an easy way to get started.

Authority

RACI does a good job of clarifying and assigning responsibility and accountability, but one of the aspect that is missing is authority. Clear definition and limits of authority is the other pillar of sound governance, besides responsibility and accountability. Review each of the DNA elements and see where authority can be defined, from a business, functional, financial, or contractual standpoint. For example, in the strategy realm, what is the project manager's or PMO's authority to kill a project if does not meet alignment criteria; from an execution standpoint, what is the PMOs authority versus the functional managers to assign resources; from a governance contractual standpoint, what is the PMO's authority to approve contract (e.g., they can approve contracts less than $50,000 only).

Decision Rights

Popularized by consulting firms like Bain and McKinsey, decision rights are promoted as a tool organize their decision-making by setting clear roles and accountabilities and by giving all those involved a sense of ownership of decisions. Clear decision rights provide a common vocabulary that allows one to cut through the confusion, by ensuring that critical decisions are made promptly and result in effective actions. In Bain & Company's version, there are five types of decision rights that are assigned to different people.

1. **Recommenders** recommend a decision or action based on assessment of data and facts, and inputs from appropriate parties.
2. **Agreers** formally approve a recommendation and can delay it if more work is required. This role may be optional depending on the organization's culture and power structures. In a decentralized or consensus-driven organization, the agree role may be held by several people whose agreement is required. To use a sales analogy, the agree role has "No" power but not "Yes" power. Limit agree roles as much as possible because they tend to slow progress.
3. **Performers** are accountable for making a decision happen once it's been made. To paraphrase Peter Drucker, decisions are worthless if there is no follow-up work assigned and completed. This role adds accountability for action for the decision. The performer role is held responsible for action, even if they subsequently seek input from others or delegate certain tasks.
4. **Input** provide recommendations based on subject matter expertise. The input role works closely with the recommender role and should be assigned only to those with knowledge, experience, or access to resources that are so important

for a good decision that it would be irresponsible for the decision maker not to seek their input. The PMO may play an input role in many decisions related to projects, programs, and portfolios.

5. **Deciders** make the ultimate decision and commit the organization to action. Each decision should have only one decider with single-point accountability.

Rules/Guidelines

Establishing simple rules can help provide the framework to steer in the right direction in complex situations. Donald Sull and Kathleen Eisenhardt, in their book *Simple Rules: How to Thrive in a Complex World*, provide a simple classification of decision-making rules that we have adopted to establish governance rules and guidelines:

Boundary rules help to decide between two mutually exclusive alternatives. Boundary rules also help you to pick which opportunities to pursue and which to reject when faced with a large number of alternatives. For example, if a business case for a project does not list at least an 8 percent or higher net present value (NPV), we will not consider it.

Prioritizing rules rank options to decide which options will receive limited resources. Prioritizing rules are particularly useful when you lack sufficient resources or time to do everything, or when people hold conflicting views about what to do. For example, it helps to have and established criteria for project ranking and prioritization based on rewards, risk, resources, strategic fit, and meeting business objectives.

Stopping rules dictate when to reverse a decision or stop or kill a project. These are probably the most needed and least practiced rules.

Review the governance activities in the DNA areas and assess which ones might benefit from establishing any of the above rules or guidelines.

Figure 6.3 provides an example of how to use the governance strands to define governance for alignment, which is related to the strategy element of the DNA.

Figure 6.4 provides a template to define governance for each of the DNA elements. The list under each of the elements is an illustrative example, and other activities can be added as appropriate for context.

As you review the governance framework, it might look like a lot and overwhelming, but view it as a menu or a governance catalog—you are not going to have everything defined from the get-go. Also, all of the strands may not apply for each of the DNA elements. Start by picking the area with the most pain and identify what are the minimum effective governance criteria that need to be defined. The idea is that as you make progress and have governance clarity for more elements, you will enhance governance intelligence over time.

Governance Strand

DNA Element	Steering Body	Standards	Policies / Procedures	Gates	Review / Audit	Compliance	Responsibility & Accountability (RACI)	Authority	Decision Rights	Rules / Guidelines
Strategy - Project Alignment & Prioritization	Portfolio Steering Comm.	PMI Portfolio Mgt. Std.	Business Case PMO Project Ranking Process	Qtly. Portfolio Review Meetings	Not Applicable	Mandatory compliance projects Top priority	Portfolio Comm.[A] PMO[R] Executives [CI]	Portfolio Comm. – Prioritize & Discontinue Above 500K needs executive approval	Funct. Mgrs [Recomm./ Input] Executives [Agree] Portfolio Comm. [Decide] PMO [Input / Perform]	Only projects greater than 8% NPV to be considered Projects not meeting at least 5% benefit req. to be discont.

Figure 6.3   Governance Framework Example

Governance Strand										
DNA Element	Steering Body	Standards	Policies / Procedures	Gates	Review/ Audit	Compliance	Responsibility & Accountability (RACI)	Authority	Decision Rights	Rules / Guidelines
Strategy										
□ Alignment										
□ Selection										
□ Prioritization										
□ Funding										
□ Resource allocation										
□ Benefits										
□ Value										
Execution										
□ People										
□ Process										
□ Technology										
□ Flow (Depend.)										
□ Risk										
Governance										
□ Effectiveness										
Connect										
□ Communication										
□ Stakeholder/ Customer Engagement										

(Continued)

Figure 6.4   Governance Framework Template

131

Governance Strand

DNA Element	Steering Body	Standards	Policies / Procedures	Gates	Review/ Audit	Compliance	Responsibility & Accountability (RACI)	Authority	Decision Rights	Rules / Guidelines
☐ Interface/ Interdep. Mgt. Integration										
Measure										
☐ Performance Reporting										
☐ Measures/ Metrics										
☐ Resource Optimization										
☐ Investment Optimization										
☐ Benefits & Value										
Change										
☐ Configuration Mgt.										
☐ Change Readiness										
☐ Change Advisory Board Adoption										
Learn										
☐ Documentation										
☐ Capturing LL Sharing LL										
☐ Learning from Failure										

Figure 6.5 (*Continued*)

PMO ROLE IN GOVERNANCE

The PMO can take a facilitative role to define, organize, and support the various aspects of governance described above. It can take a leadership role to connect and integrate governance across organizational silos, projects, programs, and PMOs. Some of the activities may include:

- Supporting and facilitating governing bodies
- Coordination and integration of governance processes
- Developing and maintaining PM standards
- Facilitating and supporting gate reviews
- Reviewing project/program performance and providing guidance
- Conducting project/program reviews
- Supporting preparation for project/program audits
- Assessing and escalating risks and issues
- Reviewing legal, regulatory, financial, industry, environmental, and other compliance
- Reviewing an improving effectiveness of governance processes

ADAPTIVE GOVERNANCE

> Putting a governor on your engine that stops the car from going over fifty-five; you're far less likely to get into a lethal crash, but you won't be setting any land speed records either.
>
> *Eric Barker,* Barking Up the Wrong Tree

Governance and agility together can seem like an oxymoron that does not belong together. But applied in the right balance, governance can galvanize strategy execution. It makes it possible to travel the highways of strategy execution in clearly marked lanes at high speeds while minimizing risk and enhancing performance. The challenge is to know how much governance is apt before it becomes like too many speed-breakers that not only slow you down but also ruin your vehicle. Chapter 2 discussed how to find the sweet spot of rigor without rigidity, appropriate for your business and organization. To design adaptive governance for agility, here we will focus on four practices: establishing simple rules, self-regulating behavior, identifying bottlenecks, and steering and making adjustments.

Rules are necessary to bring order in a chaotic environment. As described above they are one of the strands of governance. But most often they are misconstrued and misapplied based on traditional thinking. In the mechanical mindset, the belief is that the messier and more complex the situation, the more rules and heavy governance are necessary. This approach is understandable but flawed. It is the opposite, in DANCE project and program environments, too many rules can be paralyzing and confusing and lead to unpredictable and unintended consequences with increasing complexity.

The DANCE is characteristic of too many moving parts, which interact with one another in many ways, which makes it hard to envision all possible outcomes. The key is in identifying the few simple rules that can have the most impact.

Research based on complexity science suggests that complex adaptive systems all thrive based on underlying simple rules. For example, birds in the sky can deal with immense complexity and fly in formation, without colliding with each other, based on three simple rules: separation—avoid crowding neighbors; alignment: steer toward average heading of neighbors; cohesion—steer toward average position of neighbors. Similarly, ant colonies can synchronize their work and create multilayered structures without any project managers or long meetings, based on simple rules, which can be summarized as always lay a trail of pheromone and always follow the trail. Think about what simple rules could provide focus and clarity in the midst of DANCE and complexity in your environment.

Review governance processes and explore ways to make them self-regulated and nonthreatening. Just as speed indicator displays (discussed in Chapter 2) are designed to regulate traffic to a preset limit, governance processes can be used to define the boundaries with preset triggers for escalation. For example, rather than setting a project review or escalation meeting, the PMO can design and implement preset triggers that provide timely feedback to self-regulate behavior in desired ways.

Governance bottlenecks can hamper the flow of execution when there are too many layers of management decision making. One way to identify these bottlenecks is to classify and separate the different types of decisions. A good analogy is to think of a tree; the roots are the critical decisions that are best handled by key executives, on the other end are the leaves, which are the day-to-day decisions on projects in the field that should be delegated to the PMs and project teams with clear accountability. A good rule of thumb is to have centralized decision making for the root decisions that are infrequent, not time critical, and have global impact. And decentralize decisions to the leaves, when they are frequent, time critical, and do not have global impact. The tricky ones are the connecting branches that are spread and siloed and require dialog and collaboration, which hinders agility. The PMO can be like the trunk that provides stability and connects various parts and identifies and bridges any gaps or entanglements.

Remember the scenario from the beginning of the chapter, where lack of good governance was identified as a major pain-point. We worked with the PMO to prioritize some of the governance strands that could have a lasting impact. It was not easy with the different silos and personalities involved, but within 16 months the organization went from 60 percent project success to over 85 percent success. There was a considerable drop in customer complaints, and there were zero product recalls in that time frame. They started with changing the whole focus of the PMO based on the application of simple rules of prioritization based on customer pain-points, connection, and simplicity. They were successful in repurposing their steering committees with clear responsibility and decision rights. They defined clear gate decision criteria similar to the one listed in Figure 6.2. They streamlined meetings and communications, with clear purpose, agenda, and expected outcomes to make meetings and decision making effective.

Previously, a product update based on a customer complaint had to go through a circuitous approval process of four or five different committees that could take upwards of three to four months. Now, particularly if it was a customer issue, it was given the highest priority. Authority and decision making were delegated to where there was first-hand information to resolve it, cutting down the resolution time to a matter of a few days. The PMO changed from a heads-down, inside-out focus, to an outside-in, customer perspective with a greater spirit of collaboration and trust, and frequent review and adaptive governance.

The traditional view of governance is based on a defined process that has the advantage of being predictable and providing control. But its weakness is that it can be inflexible and not work well in DANCE and complex environments. The need for agility requires an empirical process that is based on observation and is adaptive to uncertain and changing conditions. Any governance framework should be dynamic and adaptive to the evolving conditions and organization's needs.

Governance tends to lean on monitoring, control, and oversight. In our practice, we view it more as a steering function. Strategic-execution through "steering" is the untold benefit of governance. The etymology of govern includes this concept from the Middle Ages: "to steer or pilot a ship." In this sense, we can think of governance as a continuous stream of course corrections to reach our objectives. Changes and "disruptions" are to be expected just as a ship captain expects to encounter rough seas from time to time. Highly effective governance means staying on course and achieving your strategy. Poor governance means getting lost, wasting resources, and frustrating stakeholders; it doesn't have to be that way.

DEVELOPING GOVERNANCE INTELLIGENCE

Use the following checklist of questions to reflect and develop governance intelligence. Pick and prioritize what to focus on over a period of time to enhance governance intelligence:

- How can we review, assess, and improve the effectiveness of our current governance structure?
- How can we improve the effectiveness of our steering bodies? Do they have a clear charter and responsibility?
- What governance actions can ensure better alignment with business objectives and strategy?
- What governance actions can ensure we are focused on the right priorities?
- How can we enhance governance for effective performance management?
- How can we implement or improve governance to ensure benefits realization and business value?
- How can we review and assess if standard methods and practices are applied, working, and effective?
- Are current governance policies effectively communicated and understood?
- Do we have appropriate policies and procedures that drive the right behaviors?

- How can we improve the effectiveness of phase gates?
- Do we have adequate project review and audit process in place?
- How can we ensure responsible use of resources?
- How can we improve the clarity of responsibility, accountability, and authority for critical activities?
- Do we have well-defined decision rights and decision rules to speed decision-making?
- How can we streamline governance processes to increase decision speed?
- How can we improve governance for proactive risk management?
- How can we ensure we are not getting in trouble?
- How can we ensure compliance to regulations?
- How can we ensure ethical conduct of project staff and teams?
- What level of redundancies and overlaps exist in existing governance structures?
- How can we remove governance bottlenecks?
- How can we better apply and practice rigor without rigidity?
- Are we tapping into opportunities for self-regulated governance where applicable?
- How are our governance processes slowing us down from conducting business?
- Is our governance structure impeding innovation?
- How can we better learn and adapt governance in the face of rapid change?

KEY TAKEAWAYS

- The purpose of governance should not primarily be oversight, monitoring, control, and limitations. Instead, it should galvanize strategy execution by ensuring alignment, proactive risk management, better performance, and decision making.
- A sound governance structure rests on the clear definition and understanding of the mutually supporting strands of governance—steering bodies, standards, policies and procedures, gates, reviews/audits, compliance, responsibility and accountability, decision rights, and rules/guidelines.
- The PMO can take a facilitative role to define, organize, and support various aspects of governance. It can take a leadership role to connect and integrate governance across organizational silos, projects, programs, and PMOs.
- Any governance framework should be dynamic and adaptive to the evolving conditions and organization's needs.
- To design adaptive governance for agility, establish simple rules, find ways to self-regulate behavior, identify and redesign decision-making bottlenecks, and focus on steering and making adjustments.

7
CONNECT

"Everything is connected ... no one thing can change by itself."

Paul Hawken

"Creativity is the power to connect the seemingly unconnected."

William Plomer

Leading Questions

- How can you reduce disconnects and better connect to catalyze strategy execution?
- What are the key DNA strands of connect?
- How can project managers and project management offices (PMOs) identify, prioritize, and link stakeholder networks and relationships?
- How can project managers and PMOs build bridges and connect the silos?
- How can the PMO connect, align, and communicate key business activities and organizational portfolio priorities?
- How can project managers and PMOs facilitate communication and relationships?
- How can the PMO cultivate a rich platform for communication and collaboration?
- Why are marketing and communication skills, rather than just communications skills, important?
- How can the PMO cultivate community and collaboration to foster connection and relationship?
- How do we develop connect intelligence?

A nearly $4 billion company I worked with decided to reorganize itself into three business units. Its PMO, which had been created a few years earlier, reinvented itself to meet a new mission of serving the three business units. Its capabilities were fairly broad, including investment planning, resource management, project, program and portfolio management tools and processes, and change management. A new director was brought in to oversee the PMO's reinvention.

It was a rocky process. The PMO's staff found itself duplicating work as it partnered with business units that weren't talking to each other. It struggled to deliver value on a day-to-day basis. Various pain points emerged in the PMO, including lack of priorities (or conflicting priorities), too-slow decision making, lack of engagement on the part of business units, and stakeholders who were unclear about the purpose and role of the PMO. There were also multiple dependencies (some hidden) across projects and programs it supported, and something familiar to any project manager: scope creep due to stakeholders who wanted different things.

There's a theme in all this: disconnection. As I talked with the PMO's leaders in-depth about their challenges, it became clear that they felt cut off from other parts of the organization in myriad ways. It wasn't just that teams within the newly formed business units didn't understand the PMO's purpose. PMO leaders felt strategically adrift: it wasn't obvious what they should be prioritizing and how the organization's strategy might translate into priorities.

To his credit, the new PMO director didn't blame others for the situation. Eventually, he realized that it was incumbent on the PMO to create the connections necessary to deliver on its mission. It could be the change agent that fixes the broken mirror, allowing everyone across the company's business units to feel part of the same picture. The director said it best:

> We were so trapped in our own bubble, thinking we were focusing on the right things and then becoming frustrated with others when we felt pushback. We didn't realize that a core part of making a PMO work is to connect. You can do everything well—PM training, resource management support, whatever—but all those capabilities end up worthless if they're not offered through strong connections built on a shared vision and trust. That's what makes what we do come alive, that's how the PMO must deliver true value.

His eureka moment came after he stepped back from the PMO's day-to-day frustrations to take a holistic view of the problem at hand. In a sense, he was stepping back to focus on the DNA of strategy execution. Realizing that they could act to repair the mirror, allowing teams across the company to see their work in the proper context and with strategic scope, the PMO team began to look for opportunities to identify disconnects across the company landscape where the PMO could build bridges, bust silos, and connect all elements of the DNA of strategy execution.

It turned out there was plenty of work to be done. The team committed itself to:

Focusing on building relationships and partnerships across the company, and at all levels.
Conducting stakeholder network analysis to flag disconnects.
Leveraging the company's matrixed structure, rather becoming tangled in it.
Reassessing who the PMO's customers are.
Better alignment around common goals.
Communicating more strategically (and deeply) across different channels.
Developing a strong marketing-communications plan.
Becoming the go-to entity for connecting gaps across projects.

These are ambitious, far-reaching goals. The PMO leaders decided they had to achieve them to truly connect the organization and drive strategy execution.

DISCONNECTED AND ADRIFT

This story captures the persistent problem we see in so many PMOs: they are disconnected from the rest of the organization in fundamental ways, and those disconnections are obstacles to effective strategy execution. At so many organizations we work with, PMO leaders feel dispirited. They don't doubt the PMO can bring real value to the table through things like supporting project and program best practices, strategic alignment, and resource management. They care about organizational strategy and believe the PMO can and should be uniquely positioned to turn the strategy into reality. Yet they know the PMO, and the overall organization isn't reaching its full potential. They're frustrated—deeply frustrated by a sense of drift and powerlessness.

So what's causing all this angst? It has to do with the problem with which this book started: the broken mirror. There are too many disconnects throughout organizations. One business unit doesn't know how to talk to another or just doesn't bother. Incoming (or long-standing) executives don't understand what PMOs do and why it matters. Employees across the organization don't understand the strategy, or how it is changing to meet new challenges. Project teams resent required documentation, from making the business case to regular reporting. People are pushing and pulling in different directions, but no one seems to be moving. They lack connection.

The problem of disconnects commonly ensnares project managers as well, whether or not they're attached to a PMO. Just like PMO leaders, project managers must grapple with stakeholder disconnects. They might be internal to the organization, involving the project sponsor or team members pulled from different departments, or they might be external, involving vendors or the customer. Regardless of the particulars of their situation, frustrated project managers will find plenty of value throughout this chapter. They too struggle to stay sane and productive in spite of the broken mirror. They're trying to align varied stakeholders around a common vision and goal to make a strategic change.

At a deeper level, what often drives disconnects is the reality of modern matrixed organizations. Matrices—think of overlapping organizational webs and cross-functional teams—in theory, should connect people in different business units.

In reality, matrices are what often cracks the mirror. "Dotted line" relationships end up feeling convoluted, chaotic, and frustrating. The PMO leaders in the story above realized that to fulfill their potential, they had to solve some of the problems created by matrices. They also had to understand the dynamics of connecting in a complex environment and an increasingly DANCE-world.

POWER OF CONNECTIONS

Connection is the foundational circulation that breathes life and adds complexity. Just like you cannot connect and communicate without a good Wi-Fi connection similarly, you cannot get things done without connection. As we discussed in the execution chapter, effective execution depends on flow, and flow can be either enhanced or impeded by the quality of the connection.

We live in a hyperconnected world where we are not only more interdependent, but also it is much easier to influence and be influenced quickly. The multiplicity of connections can add exponential complexity and intensify the DANCE. You may think that you know some of the individuals and can predict their behavior, but the reality is it is not possible to predict the behavior of the group in a complex environment. This is known as an 'emergent' property of complex systems, by which a perfect understanding of the part, does not automatically convey a perfect understanding of the whole. Connecting skills are vital to understand these dynamics and decipher the interactions and their interplay that cause complexity. Project managers and PMOs must develop skills to identify the disconnects and opportunities to bridge and strengthen the connections between the elements of the DNA of strategy execution.

The question is how can the PMO be the connective tissue of the organization that aligns the organization to catalyze strategy execution?

As anyone who works in one knows, PMOs have a unique vantage point as they interact with various business functions and units. Project managers and PMOs must constantly be looking for opportunities to identify disconnects throughout the organization and reduce them by connecting people, processes, and products to projects, programs, and portfolios, along with any other fractured organizational aspects. While integrating cross-functional activities across the organization, the PMO can be the connector and the router of information that makes the full strategic context a lived reality for employees who otherwise have a hard time seeing the big picture. It can also identify and coordinate interfaces and interdependencies across projects and programs, adding value by highlighting linkages. Finally, PMOs can help one team understand that other teams need resources—providing such context can go a long way toward smoothing over frustration and maintaining the flow of execution.

Project managers and PMOs must understand their complex environment: the interplay of multiple variables of actors, information, and interactions. Surviving in a DANCE-world requires connecting, communicating, and collaborating as opposed to reliance on command-and-control. The fundamental question that must be addressed is, how can we develop better connection capabilities?

THE DNA STRANDS OF CONNECT: HOW AND WHAT TO CONNECT

Figure 7.1 lists what to connect—stakeholders/customers, silos, business, interfaces, and interdependencies—with the how—networks and connections, marketing communications (marcom), relationships, and community and collaboration.

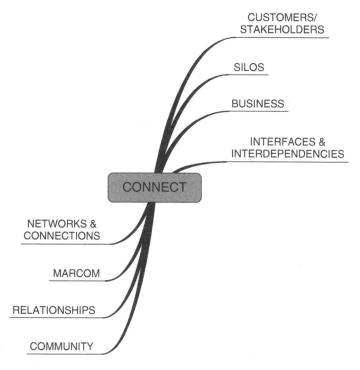

Figure 7.1: The DNA Strands of Connect

Connecting Customers, Stakeholders, Networks, and Connections (Identifying the Invisible)

In the story above, the PMO's eureka moment was the realization that without deep connections across the organization, all its passion and expertise wouldn't—couldn't—have deep and lasting impact. Put another way, it's not *what* you know, but *who* you know. And who you connect with.

Can you think of people who can get things done or have been promoted because they have connections, and not necessarily for their qualifications? A team at the Norwegian School of Business and Economics researched a similar question a few years ago. They investigated how human capital (knowledge) and social capital (relationships) contribute to individual productivity in project environments. Both are important, the researchers found; however, the social capital or who you know has a noticeably greater impact on productivity in projects.

The importance of stakeholder management in project management is well understood, and classic stakeholder management techniques like stakeholder identification, prioritization, and management based on power, interest, influence, and other factors are common. But this is not enough; the next generation of project management must go beyond and focus on networks and connections of stakeholders as well. It is harder to do an effective stakeholder analysis as the number of stakeholders grows exponentially with the increase in number of teams, vendors, partners, customers involved in today's projects. Another challenge is it is even harder to identify the stakeholders that matter. Typically, you start with hierarchical org charts, but as we all know, the real work happens in informal networks with the key connectors, who may not necessarily have the right position or title, like Gina, in Figure 7.2, but she wields a lot of clout.

So, how do we identify the Ginas of the world?

Social scientists have used the concept of "social networks" since early in the twentieth century to connote complex sets of relationships between members of social systems. Social network analysis is the mapping and measuring of relationships and flows between people, groups, organizations, computers, or other information/knowledge processing entities, according to Valdis Krebs, a researcher, and consultant in the field of social and organizational network analysis.

As Friedrich Nietzsche beautifully put it, "Invisible threads are the strongest ties." The challenge is how to identify the invisible connections. In our practice, we have adapted social network analysis to analyze stakeholder networks, a technique that helps project managers or PMO leaders understand the complexity of their networks and the connection points that most matter. It allows us to decipher the intricate maze of relationships and their impact on complex projects. It's a variation of social network analysis, which analyzes the connections of nodes and ties in a social network. Detailed network analysis involves a combination of measures like:

Degrees—the number of direct connections.
Betweenness—how much a node controls what flows in the network.
Closeness—how quickly a node can access all other nodes via a minimum of hops.
Other factors—like diameter, density, and subgroups.

Figure 7.3 illustrates an example of a project stakeholder network. We can assume that Brian is the project manager and Debbie, Christine, Rich, Ahmed, Kumar, and Chen are core team members. Tom has some supervisory capacity over this team; Pat is the sponsor and Beth the client.

These types of visualizations are like mirrors that help you spot who are the most connected and powerful stakeholders in this network? Can you see who the key people are in this example? Can you identify any structural gaps that need to be bridged? Upon preliminary review, it may appear that Brian is the most connected and powerful person in the network. However, further analysis reveals that Tom is the key node in this network because he is the gatekeeper between the project team and the sponsor and client side of the network. Next, Debbie and Christine are vital nodes between the right and the left side of the network. Even though Brian appears to be well connected and busy with his team, he needs to bridge the gap between himself and Pat and Beth, instead of relying on Debbie and Christine or Tom. If he does not have a good relationship with either of them or if either of them decides to withhold information, Brian can be negatively impacted.

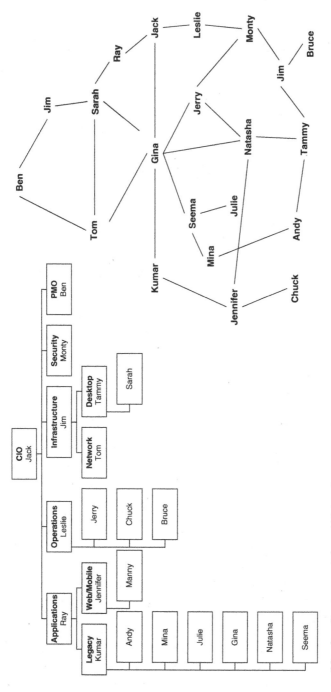

Figure 7.2   Org Chart versus Informal Network

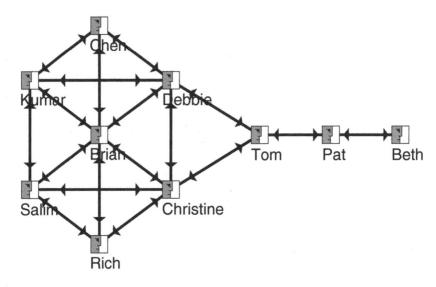

Figure 7.3: Stakeholder Network Analysis

Start by simply illustrating the network of stakeholders in your project or PMO environment and identify the key nodes and any gaps between stakeholders that might need to be bridged. You'll likely be surprised by what you find—new ways of viewing your stakeholder landscape, and the relationships that collectively contribute to (or limit) its impact. Review Figure 7.4, adapted from Mariu Moresco and Carlo Notari, to get deeper insights into the attributes of elements in informal networks. Particularly interesting are the features of the network actors. For example, you can see how Gina in the above example might be the central hub or opinion leader. In large projects or organizations, it is critical to identify the Ginas of the world who are the connectors, and the important nodes that can either spread or block the message. You might not have the bandwidth to touch or influence everybody—that's why you have to identify the powerful, influential nodes and strengthen your connection with them. This analysis also helps you to identify the gatekeepers, power brokers, boundary spanners, pulse takers, and positive deviants, so that you can leverage the right type of connectors.

Stakeholder network analysis is a powerful technique. You can leverage it in powerful ways depending on what you know about who you know. We can all relate to this living in a hyperconnected world of social media with LinkedIn, Facebook, Twitter, and other platforms that are based on the same idea of the power of connections. While LinkedIn and Facebook use powerful tools and algorithms for network analysis, there is a whole slew of tools like Inflow (which was used to create the above example) and many others. Just Google "social network analysis and visualization tools" and you will see the latest links, like Commetrix, Cuttlefish, Cytoscope, Egonet, Gephi, Netyltic, Netminer, and others.

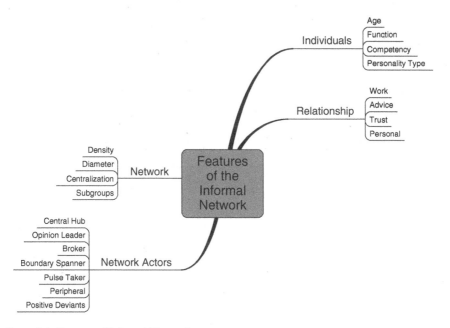

Figure 7.4: Features of Informal Networks
Source: Adapted from Mariù Moresco and Carlo Notari, "Stakeholders' Worlds" chapter in Projects and Complexity.

To start with, you don't have to use sophisticated network analysis tools, you can simply be more observant and use a stakeholder leverage matrix as shown in Figure 7.5. This matrix can be used to map the existence of a relationship among a group of people with a checkmark or X. You can add another dimension by using a scale from 1 to 5 to assess the degree or depth of connection.

If we can understand what is driving behavior, and why adversarial relationships develop in the first place, we're more likely to be able to shape behavior and deepen connections. That's why from a next-generation PMO standpoint, stakeholder mapping, and stakeholder network analysis aren't enough. We need to go deeper and dive into behavior and what's behind it. New psycho-linguistic tools are now emerging that allow one to do just that. These tools are designed to mine e-mail and various social interaction communications based on algorithms to better understand relationships and highlight behaviors like bullying, passive aggressiveness, and emotional heat maps in project teams.

It's easy to see the value in understanding certain stakeholders with such depth. Ultimately, though, it's not about *what* you know about *who* you know, but *what you do* with *the what* and *the who*, that is going to matter.

Connecting Silos: Bridging the Matrix

"One hand doesn't talk to the other." "The silo mentality." Sound familiar? If you work in a large matrixed organization, you are dealing with it daily. Organizations

Stakeholder Leverage Matrix										
Key Stakeholder	Who	They	Connect	with	&	Influence				
	1	2	3	4	5	6	7	8	9	10
1	X									
2		X								
3			X							
4				X						
5					X					
6						X				
7							X			
8								X		
9									X	
10										X

Figure 7.5: Stakeholder Leverage Matrix

are traditionally hierarchical and siloed, with top-down, division of labor, fragmented processes and culture. Many PMOs, of course, are created to standardize processes across the organization. But the fragmentation is much broader and deeper than just processes. In my experience, it often spans strategy, core assumptions about the organization's purpose and awareness of what other departments and business units are accomplishing—or trying to.

How did things get so complicated? The matrixed structure of most large organizations is a good place to start. By design the matrix helps to further crack the mirror. Companies should know better in this day and age, but there seems to be no stopping the trend.

As Ron Ashkenas, author of *The Boundaryless Organization*, once noted in the *Harvard Business Review*, although the speed of globalization and technological innovation in the twenty-first century demands shorter decision cycle and stronger collaboration, too many companies haven't updated their org structures for the times. In fact, he writes, "having to cope with a fast-changing global economy has led many companies to create even more complex matrix organizations, where it's harder to get the right people together for fast decision-making."

The original idea for creating matrixed organizations was to create intersections "between global businesses and local resources, between technical expertise and business units and among multiple functions," as Susan Finerty puts it in her book *Master the Matrix*. Matrix organization structures first became fashionable in the 1960s in the aerospace industry, and then spread beyond in the ensuing decades. The approach has since become so common that organizations don't even necessarily use the word, *matrix* anymore—but that doesn't mean project managers and other employees aren't feeling frustrated by poorly defined roles and unclear lines of authority. Finerty writes:

> Much has been written of project managers who sit at the crossroads between reporting through the project management structure and through their "business" boss as described above. But the matrix isn't just for project managers anymore. Informal matrices and matrixed teams are cropping up in organizations that don't call themselves "matrix organizations." This once formally named and managed structure has morphed into overlapping organizational webs that are often navigated with little or no guidance. These webs are sprouting up as the traditional business structure of multiple, independently operated business units shifts to shared services, cross-functional teams and "flatter" organizations. All these efforts are aimed at doing more with less and gaining "economies of scale," and they all create matrices. In addition, automation, globalization, regulation, and legislation have created a reality in which few tasks, projects, or goals fall neatly into one person or team's bailiwick. Instead, they cut across teams, functions, and geographies. ...
>
> Matrix roles (and the challenges and frustrations that go with them) are everywhere. The field customer contact person who has to work through a maze of resources to write contracts, negotiate delivery and troubleshoot product issues. The product manager in Singapore who is simultaneously accountable for numbers in her country, region and business unit despite the fact they are in conflict. The HR person who reports to a globalized HR function, with a dotted line to the head of the business he/she supports. All of these are matrix roles. When the matrix practitioners I connected with described their matrix roles they used words like confusing, chaotic, convoluted and frustrating. Operating in a web of authority, on multiple dimensions and sometimes at cross-purposes can be draining to even the most energetic matrix managers.

It's not hard to see how organizations can become very good at producing something bad: disconnects. Let's see if the following scenario sounds familiar. A company's strategy is deployed top-down, cascading from the CEO and other executives to business silos, smaller units, and ultimately individuals. Along the way, each function develops its own metrics, often in isolation and cross-purpose from other departments working toward the same goal. Silos are thus reinforced, and dysfunction rears its head.

Next-generation project managers and PMOs have to learn to master the matrix and bridge organizational silos. The goal is a panoramic vision, and strategic mindset shared across the organization. The question is, how is this mindset developed exactly?

Connecting Business Activities and Organizational Priorities

To not only survive but thrive, every business must have a uniquely powerful strategy. But strategy means nothing without the right execution. And execution isn't possible if the strategy and the underlying business activities are misunderstood or ignored.

That is why one of the elements of the DNA of strategy execution is to connect, clarify, and align the strategy and business activities with the projects and PMOs.

A PMO dedicated to nothing but best project management practices and processes is a strategic opportunity wasted. Of course, many PMOs go beyond and practice portfolio management to align and prioritize projects and programs to the organization's strategy.

That's great. But the next generation PMO must go beyond and must be inspired by and infused with the organization's strategy. All of its activities must be unabashedly dedicated to advancing awareness of the strategy and ensuring its execution. This is its raison d'etre: to link projects and programs to the organization's portfolio priorities, and its broader business context and activities.

Large matrixed organizations with far-flung offices and layers of middle management are seemingly designed to help people lose perspective. Teams become obsessed with putting points on the board, even if they're playing the wrong game. Key performance metrics may be achieved but not move the needle on strategy execution. Meanwhile, what really matters—customer experience and ultimately customer satisfaction—gets lost.

Businesses have always been built around customers, but because of the Copernican revolution in management discussed earlier, companies are increasingly competing to better understand and serve customers. The PMO has an important role to play here: in its many interactions with various parts of the organization, it can hammer home what every employee ought to have memorized: how the company's business model works and how it creates value for customers. The business model canvas (BMC), described in Chapter 4, is a great way to concisely and visually highlight all the components of a company's business model. It features nine elements covering everything from value proposition to cost structures to customer segments. You can use the output of the BMC and plugit into the business alignment matrix, also discussed in Chapter 4 to show how the project or PMO activities are aligned with the business and stakeholders.

Next-generation project managers and PMOs ought to have an understanding of all business model components because that opens opportunities for them to add value. They can see key business value-creating activities, and the channels through which it reaches different customer segments. Or fails to.

This last piece is critically important. PMOs have an opportunity to facilitate an organization-wide understanding of the customer's perspective. It is human nature to only think from one's own perspective, or the perspective of your tribe, team or business unit. But that's a recipe for complacency—which is what smart and agile companies try to fight against every day. Because by their very nature they intersect with all parts of an organization, PMOs are well positioned as a key ally in the fight to win in a customer-centric world. Strategies are more and more customer focused—which means execution can only be successful if it keeps the customer's perspective in mind from start to finish.

See the Gap and Make the Connection

To work in a large matrixed organization is to, at some inevitable point, feel lost. PMOs can help people find what they're looking for. Here's what I mean: remember the PMO in the $4 billion company I described at the start of this chapter? While I was working with its leaders, one of them told me an instructive anecdote.

The company is based in the United States but has offices around the world. One day a project manager in Australia e-mailed the American PMO leader seeking help: He was looking to find a few people with certain specialized expertise to bring into his project team but had no idea how to find such people. (Ideally, they'd also be in Australia.)

Given the PMO's broad experience partnering with various departments and business units across the organization, the PMO leader had no trouble identifying staff in Australia that could be a good fit for the project manager's project. After connecting him with the individuals, he couldn't help but notice the organizational and geographic absurdity of the situation: the project manager had to reach halfway around the world to find some colleagues in his backyard.

Such is the reality of many global corporations today. It's a reality that smart PMOs can easily improve by proactively identifying structural holes in the organization. By linking and integrating otherwise disconnected silos, they can prove how their unique role drives value. The PMO should always be anticipating the sort of problems the Australian project manager needed help solving, and then creating and strengthening connections between different parts of the organization.

In a nutshell, the PMO should build the bridges that should have been built long ago.

Connecting Interfaces and Interdependencies

We all wish that we could work by ourselves and get things done, like craftspeople and artists, but project work by nature is interdependent. What makes it challenging is when you find out about a project underway that is, unbeknownst to the project team, reliant on another project's progress or success for its own progress. It's not just that the left-hand doesn't know what the right-hand is doing—it's that they're on a collision course. This is a common challenge as various parties in projects are loosely coupled, whereas the tasks themselves are tightly coupled and interdependent as observed in an MIT article, "What Successful Project Managers Do." When unexpected (DANCE) events affect one task, many other interdependent tasks are also affected. Yet the direct responsibility for these tasks is distributed among various loosely coupled parties, who are unable to coordinate their actions and provide a timely response. Project success, therefore, requires interdependence, as well as an understanding of the related interfaces.

For each of the interface and interdependency, design appropriate people, process or tool resolution strategy that can enhance the flow. Overall the PMO can promote a culture of interface and interdependency awareness and focus on flow, and look for opportunities to remove any blockages that impact the flow and impede execution as discussed in Chapter 5.

Because project managers and PMOs reach across the organization, touching all kinds of projects and programs in the course of its activities, they can highlight interfaces and interdependencies that might be invisible to those in the trenches. They can illuminate all that the organization is doing to those with obstructed views—which is most people. Various good results can follow: interdependent projects can be better coordinated, expert resources can be efficiently shared, project redundancies can be avoided, and frustration about why certain teams can't obtain necessary resources to executive high-priority projects can be alleviated.

In a perfect world, PMOs wouldn't be needed to play this integrative role: the organization would be so in tune with itself through the course of normal business activities that everyone would understand their full context.

Holacracy: Antidote to the Matrix?

In the age of agile, traditional organizational structures and the hierarchies and matrices they generate, are being challenged, and companies are trying radical new approaches on their tilt toward agility. One of the concepts is holacracy, made famous initially by Zappos, and online retailer, now owned by Amazon. Holacracy is an "operating system" for self-management in organizations. It is based on the ideas of application of principles of complex adaptive systems. Instead of traditional hierarchies and job descriptions that might be revised every few years, the system "allows businesses to distribute authority, empowering all employees to take a leadership role and make meaningful decisions," according to Holacracy.org. Roles are regularly updated to align with changing needs, teams self-organize, decisions are made locally, and all are bound by the same rules, from the CEO on down.

This, at least, is how holacracy is supposed to work. Zappos, the online shoe and clothing store, is the most prominent company to have adopted the system wholesale. It started transitioning to the system in late 2014, and although about 6 percent of employees cited holacracy as the reason they were taking severance packages in 2015, CEO Tony Hsieh, a big proponent of holacracy, wanted to build a more self-directed culture that gives employees a stronger sense of purpose—and unlocks innovation. He wrote:

> Research shows that every time the size of a city doubles, innovation or productivity per resident increases by 15 percent. But when companies get bigger, innovation or productivity per employee generally goes down. So, we're trying to figure out how to structure Zappos more like a city, and less like a bureaucratic corporation.

Other organizations have tried holacracy but ultimately abandoned the system. The social media company Medium, for example, spent a few years operating through the system, but in March 2016 announced it was "moving beyond" the system because it was "difficult to coordinate efforts at scale." Holacracy's "deep commitment to record-keeping and governance" ultimately hindered a proactive attitude and sense of communal ownership," wrote Medium Head of Operations Andy Doyle. But the management model most companies use is more than 100 years old—and Doyle said his company is determined to operate in ways to recognize the speed with which information flows today and the diversity of team members' talents.

Connecting Is Communicating

The effectiveness of connection depends on the quality of communications that flows through the connections. Communication is the lifeblood that flows through the organization and the survival of the organization, or for that matter project managers of PMOs depends on the quality of communication. For the most part, the level of communication is typically superficial and procedural—what to communicate, with whom, using what medium and with what frequency. What they rarely do is focus on the quality of communications and the behavioral dimensions of the stakeholder landscape. The key questions you need to address are: *How do you connect, communicate, and build relationships and rapport with executives and key stakeholders at multiple levels? Particularly in today's turbulent world where everyone is drowning in multiplicity of communications, how do you get your stakeholders attention? How do you separate the noise from the signal and focus on what's important in a clear, concise, and direct manner that resonates? How do you navigate the politics and hidden agendas? How do you create a rich environment for straight talk based on trust, mutual respect, and collaboration?*

Earlier in this chapter, I touched on the importance of understanding what's behind behaviors detrimental (or even adversarial) to a PMO's effectiveness. I noted that some new tools are starting to emerge that can help teams dig deeper to understand who they're dealing with. In truth, understanding the behavioral environment surrounding the PMO requires a lot more than just tapping the power of a specific tool. For the PMO to forge deep connections throughout the organization, it needs to think of itself as almost a steward of its environment, constantly appraising, and the specific personalities and interpersonal dynamics that surround it. It's a panoramic, never-ending process.

Traditional communication approaches rely heavily on formal communication channels such as mass e-mails (e-newsletters), one-to-one e-mails, scheduled meetings, and a project Website that can be updated. These plans to discount more informal channels like impromptu hallway conversations and coffees or lunches, which are rich, spontaneous, and interactive. They help not only in communicating but also in building connections and trusting relationships. Memos, website updates, e-mails, and regularly scheduled meetings do not have the same impact. These formal channels used extensively, can have a numbing effect and lack the quality and intensity of the personal, on-the-spot interaction needed in today's world of information overload.

An argument against informal communication channels is that things can get out of control when discussions are informal, and there is no proof or documentation. Informal channels also introduce the opportunity for rumors spreading through the grapevine. While these are valid concerns, they can easily be addressed.

Follow up informal conversations and communications with a written confirmation, or vice versa.

The idea is to strike a balance between using both formal and informal communication channels for the most effective way to gather input and deliver your message. Here are some tips to effectively balance the use of formal and informal communication channels while trying to forge connections throughout the PMO:

Switch. Typically, formal communication channels are used first, and when there is no response, then you are compelled to use other informal means. A more effective way might be to use informal means first and then follow-up with a formal confirmation. This way you have planted the seed; people are expecting your communication and may be more responsive.

Seek feedback. Build ongoing feedback loops to assess whether you are using the right communication channels. Informal means are more effective to measure and gain insights about stakeholder satisfaction. Some people may not be comfortable speaking up in formal meetings, but they are more open in informal settings.

Formal communication is like a project's skeleton, providing structure. Informal channels are the nervous system that provides a network to facilitate communication. Both are necessary for navigating the matrix.

Why Communication Is Not Enough—You Need a Marcom Strategy

In today's world, it is not enough to have a communication plan; you need a marcom approach—marketing and communications strategy—to market, brand, and promote your initiative, project, program, or PMO. To overcome any negative perception, a marcom strategy is key to continuously have a pulse of the perception and manage expectations with appropriate messaging and communication.

Here are some useful tips to keep in mind as you, develop a marcom plan for your project, program, or PMO.

- **Plan your marcom** by focusing on the objective (what you want to achieve), message (what you want stakeholders to know), and media (which communication vehicles best convey the message). You can also use marketing models such as AIDA for creating awareness, interest, desire, and action.
- **Create a compelling message** that will resonate with your target stakeholders. Be sure to address an immediate need or issue and focus on benefits. Don't be afraid to brand your PMO with a motto, such as *Where strategy becomes a reality*, or *Connecting the C-Suite to the Trenches*.
- **Develop elevator (60-second) and water cooler (3- to 5-minute) speeches.** Train your project team so that everybody communicates crisp and consistent message.
- **Explore the use of social media tools.** Tweet your project, establish collaborative project wikis or blogs or podcast your sponsor or key stakeholders. (More on this below.)

- **Be careful of overhype.** Don't create too much buzz, oversell, or set too-high expectations that can backfire.
- **Iterate.** Continue marcom plan activities on an ongoing basis. Tailor timely messages to specific target audiences, but stay consistent with the central theme and goals of the project.

PMO leaders who create and enact a comprehensive marcom plan will gain a distinct advantage and see their PMO receive greater buy-in, implementation support, and overall acceptance across the organization.

Leveraging Social Media

Does your project or PMO have a social media strategy? Social media channels like LinkedIn, Facebook, Twitter, and many others have redefined our identities as friends, employees, and customers. These channels are ubiquitous and can be the initial step to get started with stakeholder and network analysis, just by Googling someone or glancing at their LinkedIn connections. To optimize their profile, relationships, and impact, PMOs must meet people where they so often are. They need to understand the power of particular social media channels and leverage them to understand and build their networks.

Social media present more than just relationship-building opportunities, however. There's also huge potential for marcom, branding, and making your PMO's purpose and value proposition clear, as well as propagating viral ideas and content. Social media is also a great way to monitor various networks that ultimately determine the PMO's impact. For example, what are customers saying about the company's products and processes? Understanding what's working and what isn't ties directly into the PMO's ability to position itself as a crucial strategy execution vehicle. When the PMO can understand exactly how past projects have failed to deliver promised benefits, it can better assess the strategic promise of proposed or midstream projects—and potentially speak up to push for better alignment. Social media can also reveal what various internal stakeholders are saying about the PMO's processes or projects. Often, you can't afford to miss such feedback—if it's outside of normal company communication channels, it could very well be more candid than otherwise. Unvarnished opinions may sound harsh or feel unduly personal, but they need to be heard. Even if it's just a PMO perception problem, that's still a real problem. Tapping social media can be critical to uncover certain challenges the PMO faces in connecting across the organization and delivering on its promise.

Relationships: Strengthening Relationships and Developing Partnerships

Effective organizational project management and PMOs derives strength from the support of others and evolves on the basis of interpersonal relationships. It must be actively engaged in building and managing the network of relationships across the board. Healthy relationships build a rich platform for communication and collaboration. The more relationships we develop, the more potential partnerships and a rich environment for straight talk based on trust, mutual respect, and collaboration.

In a matrixed world of work, relationships are the lubrication that gets things done. Without them, we'd all just be confused cogs in a wheel that won't turn. Here's another pertinent passage from Susan Finerty's book *Master the Matrix*:

> Job titles no longer bring with them everything needed to get the results that we are held accountable for. "Decision-maker" distinctions that used to be illustrated by title and office size are disappearing. Pinpointing who is in the catbird seat becomes less and less clear, and multiple customer groups with disparate needs tapping into the same pool of resources drown out the traditional cry of "the customer is always right." ... Getting results in matrix roles is as different from traditional roles as basketball is from swimming. Different rules apply; different skills are needed.

The all-important question is, what kind of relationships should you build, and where and when? Part of the answer here is somewhat straightforward: you want to target the people best positioned to lubricate the wheels that push a project or program forward to the finish line. That could mean external stakeholders (such as vendors or partner organizations) or internal stakeholders (such as a project's executive sponsor or team leader.)

What ideally should come out of all these relationships is a clear sense of partnership. Everyone should be crystal clear on what the common goals are, and have total clarity on the extent and responsibilities of each person's role, and respective decision-making powers. Sorting all of this out helps any PMO or project leader manage the matrix and foster connections.

But some relationships won't be so oriented toward immediate work and needs—they aren't as transactional. For example, imagine a PMO that from time to time develops training programs for project managers. It would want to do so in partnership with the organization's human resources (HR) department. Sure, the PMO leader could just knock on HR's door and make its needs clear. But HR is much more likely to respond quickly and favorably if there is a prior, ongoing relationship between it and the PMO. If the connection is already there, there's already a foundation upon which to build the new training program. The PMO doesn't have to start from scratch, and the work at hand is likely to be more enjoyable because it will partly reflect the good will and collegial sense of shared mission between the PMO and HR.

Your connections define who you are and your clout, or lack thereof. Next-generation project managers and PMOs need to constantly think about their 360-degree landscape to widen and leverage the overlaps between their operational, personal, and professional network. This key question must be asked: who else do we need to proactively develop a partnership with to be successful? Don't want until tomorrow to find the answer. You need to be connecting and building bridges today.

Community and Collaboration

Connection and networks enable community and collaboration creating a vibrant digital ecosystem, beyond the silos. If you are working separately in a silo with 1,000 people, and you have 10 ideas, each person only has 10 ideas. If you are working in a collaborative sharing community, your collective intelligence is 10k ideas each. How do you create a rich platform for sharing and a collaborative community in your project management and PMO environment? You can transform or complement your PMO to a community-oriented model, which is based on the concept of a community of practice (CoP).

A CoP is a group of people bound by common interest who are engaged in real work over a period. They build things, solve problems, and learn and invent new ways of doing things. In his book *Communities of Practice, Learning, Meaning, and Identity*, Etienne Wenger, PhD, a pioneer of CoP, defines it as a group of people who share a concern or a passion for something they do, and learn how to do it better as they interact regularly.

Many companies like Google, Apple, Caterpillar, World Bank, and others have encouraged communities. They recognize the effectiveness of informal structures to promote learning and sharing of knowledge and best practices, coupled with the convergence and popularity of social networks and the associated collaboration technologies. A community-based PMO can harness the knowledge that already exists within the organization. Instead of the PMO playing the role of an elite subject-matter expert, it facilitates a subject-matter network to propagate practices that are more relevant to the community.

How do you build a community-based PMO? Created and initiate a number of communities such as methods and processes, learning and development, measurement and metrics, and strategy alignment. Facilitate the communities and provide a collaboration platform for the communities to meet, self-organize, and improve practices.

In CoPs, the emphasis is on "practice" where practitioners practice, collaborate, and improve their craft and produce results. Each community can create its own plan. For example, the strategy alignment community could develop a project management methodology with tools and templates that were meaningful to them.

If you think that your organization is not ready to implement a full-fledged community model, you can start small. As a pilot, you can start a lunch-and-learn group of people who are interested in project management (for example) and pick a topic to discuss, work, and improve upon. As word spreads and people see results, you can grow the community organically.

Do you want to build a dreaded control-oriented PMO—or a studio or lab where everybody wants to try and experiment? A community-based PMO is an effective way to transform your project management culture and gain buy-in and acceptance of your PMO. People help create it. Why would they resist or reject it?

Don't get dejected if initially your community does not take off, or after its peak it gradually withers away. This is normal, as you see with Internet groups on different platforms, communities have their own lifecycle and resurgence, just like any social groups. To make it effectively work and sustain it needs good facilitation and leadership skills.

Leadership isn't about making people do things; it's about making people *want* to do things. PMOs should ideally be cultivating a space where people can connect with each other, solve common problems, and collaborate in creating methods that are appropriate for them. The goal is a virtuous cycle, where people volunteer to share lessons learned and influence each other to use best practices. Project management spreads from the bottom-up, and everyone collectively owns and values best practices, while also understanding their larger purpose: strategy alignment and execution.

DEVELOPING CONNECT INTELLIGENCE

Review the following questions to assess and develop connect intelligence:

- How can we actively identifying disconnects across the organization?
- What can we do to reduce the disconnects?
- How can we better connect and communicate business strategies, priorities, and activities?
- How can we assess and leverage the power of networks with stakeholder network analysis?
- How can we widen and leverage the overlaps between our operational, personal, and professional network?
- Who do we need to connect with and develop relationship with to be successful?
- How can we forge the right connections to prevent messages from being overlooked, ignored, or rejected?
- How can we build bridges and highlight links invisible to those in the trenches?
- How can we better connect interfaces and interdependencies across projects, programs, and portfolios?
- How can we design a marcom strategy to continuously set expectations and manage perceptions with appropriate messaging and communication channels?
- How can we better utilize social media and collaboration platforms and tools?
- How can we better identify and utilize the power of connectors and positive deviants?
- How can we utilize community and collaboration approaches to increase sharing and collective intelligence capabilities?
- How can we embed sensors to collect connective intelligence for effective sense, respond, adapt, and adjust (SRAA) capabilities?
- How can we cultivate deeper relationships with our key stakeholders and customers?

Connect is the element that links the other strands of the DNA of strategy execution together. You can master the other strands, but success will hinge on whether the project or PMO has the right kind of connections with the right parts of the organization, and knows how to communicate through them. Here is a variation on the famous Albert Einstein equation $E = mc^2$. While Einstein's equation is about the relationship of mass to energy, in this context we can say that the relationship of a project manager or PMO's effectiveness (E) depends on the mastery of the DNA elements and the connections and communications that flow through. In other words, you can master (m) all the other elements of the DNA, but without the right connections and communications (c^2), success will prove elusive. It all depends on the power of connections.

KEY TAKEAWAYS

- Many PMOs (and project managers) feel disconnected from the broader organization and dispirited by a sense of drift and powerlessness. This is often a by product of the disjointed reality of modern, matrixed organizations.
- Everything is interconnected and the DNA elements cannot function or evolve optimally without making the connections.
- Project managers and PMOs must understand their complex environment: the interplay of multiple variables of actors, information, and interactions. Surviving in a DANCE-world requires connecting, communicating, and collaborating as opposed to reliance on command-and-control.
- Project managers and PMOs are uniquely positioned to link project, program, and portfolio priorities to business activities and the overall strategy. They can be *the* connector that makes the strategic context a lived reality.
- Connect stakeholders, silos, business, and interfaces and interdependencies by focusing on networks and connections, marcom, relationships, and community and collaboration to develop overall connect intelligence.
- Go beyond stakeholder identification and management to stakeholder network analysis to understand the underlying stakeholder informal networks where relationships are forged, and real work gets impacted. Assess and analyze your network and find ways to increase the power of your network.
- Project managers and PMOs should constantly be engaged in building bridges that bring the entire organization into the same strategic vision. They can and should link and integrate all the parts into a whole.
- It is not enough to just communicate; you have to practice marcom, marketing, and communication, the ability to brand, position, market, and sell your project or PMO.
- Relationships lubricate all the wheels delivering project management value; strive to cultivate deeper connection and partnership with key stakeholders and customers.
- Developing connect intelligence with better stakeholder and network analysis can enhance Sense-Respond-Adapt-Adjust (SRAA) capabilities.

8
MEASURE

"Not everything that can be counted counts, and not everything that counts can be counted."

Albert Einstein

Leading Questions

■ Why is measurement elusive and challenging?
■ Are we measuring what matters?
■ What is the purpose of measurement?
■ What are the DNA strands of measurement?
■ How do we define success and how do we measure it?
■ How do we come up with the key measures that drive the right behavior?
■ How do we decipher the measure and related algorithm that leads to effective strategy execution and desired results?
■ What is value, and how do we measure project management (PM) and project management office (PMO) value?
■ How to develop measure intelligence?

Karen was focused on burndown charts and tracking the completion of user stories in each sprint as a software development project manager. Her manager kept asking her how many stories they had completed. There were high hopes that after they had adopted agile, things would be better. This time, they seemed to be on track, but at the last minute, they had to pull the plug and could not go live for a crucial feature the customer was expecting.

Kumar is under pressure to reduce the number of bugs. His manager had made it clear that all eyes are on the bugs and we need to get them down. Kumar has been working hard and putting in a lot of extra hours, but the bug list doesn't seem to shrink. He is wondering if he is destined to fix bugs for life and what he can do about it.

Ken led the PMO for over two years in a healthcare organization. He felt a sense of pride with all the PMO team's accomplishments, and he thought he had the metrics to prove it. When I talked to him, he described how they had implemented a standard project management methodology with 90 percent compliance. They had provided standardized training to 100 percent of project managers, out of which 88 percent had gotten certified. The project success rate had increased, and 85 percent of their projects were on time and budget. And yet he wondered why the executives continued to challenge and question the PMO's value and did not care about what they had accomplished.

WHY DOES MEASUREMENT CONTINUE TO BE ELUSIVE AND CHALLENGING?

The above scenarios highlight the prevalent enigma of measurement. Even though we are in the age of big data, machine learning, analytics, and algorithms and sophisticated measurement systems with key performance indicators, scorecards, and dashboards, measure remains an elusive and challenging element of the DNA of strategy execution. Today, we have the capability to measure everything. The increasing type, volume, and frequency of data streams we are involved in creating or reviewing is drowning and dumbing us into busy work, and we barely have a moment to look up from our devices and ask what's the point? How could Karen, Kumar, and Ken in the above scenarios, develop measure intelligence to come up with better measures to drive the right behavior and desired outcomes?

> "That which is measured improves, but if you are measuring the wrong thing, making it better will do little or no good."
>
> *Michael Hammer*

Management educator and MIT professor Michael Hammer summarized the challenges in an MIT Sloan Management Review article, "7 Deadly Sins of Performance Measurement," in 2007:

Vanity. Using measures that will inevitably make the organization, its people, and especially its managers look good. Just because it is being measured doesn't mean it is important. There could be a lot of other things that are important but are not being measured.

Provincialism. Siloed approach to measurement. Letting organizational boundaries dictate performance measures leads to sub-optimization and conflict. Creating competition among departments seems compelling, but strong incentives tied to strong metrics force people to concentrate on just one part of the work, neglecting other contributing factors needed to achieve a goal.

Narcissism. Measuring from one's own point of view rather than from the customer or stakeholder's perspective.

Laziness. Measure what they have always measured, rather than go through the effort of ascertaining what is truly important to measure. Project management is particularly guilty of this, as we continue to focus on triple constraint metrics because projects have always been measured that way.

Pettiness. Measuring only a small component of what matters, for example, just tracking the number of bugs instead of the overall quality system.

Inanity. Implementing metrics without considering the consequences on human behavior and performance. People learn to game and manipulate metrics. A common example is using call duration to measure the performance of customer service reps, which drives the unintended behavior to rush through calls, instead of taking time to solve the customer's issue.

Frivolity. Arguing about metrics and looking for ways to pass the blame to others rather than shouldering the responsibility for improving performance.

These measurement sins are still pervasive even with today's sophisticated measurement collection and reporting systems. In some instances, measurement malaise is worse today in a world of page views, clicks, and likes, where these measures are manipulated, misinterpreted, and drive the wrong behavior. The pursuit of meaningful measurement remains elusive because measurement is not easy—deciding what to measure, how to measure, what behaviors our measures will drive, how to analyze and interpret the data, what the unintended consequences will be, and how these measurements will be manipulated are all tough to decipher.

Measuring project/program and PMOs is fraught with its own challenges. Project measurements have been bound within the triple constraints of cost, time, and scope for a long time. The question is, do these metrics promote the right behaviors and outcomes? Or are they harmful, as the project manager is driven to rush the completion of a milestone to meet the deadline, only to find out the customer cannot use it?

Typically, measurement approaches are rooted in the machine-oriented, industrial, cause–and-effect, linear paradigm, as we have been discussing in this book. The problem is that today's project environments are more like complex ecologies in which it is hard to pin down a direct cause and effect. As a result, some of the benefits are nonlinear and unpredictable and may be hard to trace and attribute to specific initiatives.

To capture intangible value, estimates of measurements are often based on assumptions and calculation that could be questioned. Let's assume your annual budget is $100 million for projects. Planned revenue is $500 million and 50 projects are planned to be delivered. A PMO is implemented with the aim of achieving project efficiencies and increased project productivity. At the end of the year, you deliver 60 projects. The actual cost is still $100 million, but actual revenue is now $550 million, an additional $50 million in revenue! The challenge in this example is to isolate PMO value and attribute the benefit of additional revenue to the PMO. This example is based on several assumptions that could be questioned: that we could do more projects because of efficiencies brought by the PMO alone, that the basis of calculating additional revenue brought by 10 additional projects is legitimate, that the organization has a good handle on its portfolio, that the project and revenue information is based on reliable and accurate information, and so on.

What gets measured is what's easy to measure, not necessarily important. As Dan Ariely points out in his book *Payoff: The Hidden Logic That Shapes Our Motivations,* organizations overemphasize the countable dimension. Following the principle of looking for your keys under the street lamp, managers are drawn to the subset of tasks that are easily measurable. Consequently, they overemphasize those parts of the job and divert attention and effort away from the uncountable dimension. The other mistake is to treat the uncountable dimension as if it were easily countable. In fact, reducing the intangible to something simplistic and countable often misses the point, for example, measuring the number of reports produced and accessed, rather than the quality or the benefits of the report.

The difficulty of capturing intangible value and distrust of soft measures often forces organizations to measure what can be measured, rather than what should be measured. They rely more on traditional metrics that are based purely on financial or hard measures. Some organizations with a heavy financial focus ban intangible metrics from consideration. This leads to a short-sighted approach to measurement and does not capture the enabling and indirect value. Unfortunately, these measures often drive the wrong behaviors and promote gaming and adjustments to meet required targets.

A common challenge associated with selecting, collecting, and communicating measures is the ability to document, record, and capture good data, the reliability of systems, and the accuracy of the information they generate. Benefits measurement data is not easily available or imprecisely defined and collected and communicated haphazardly. Some organizations with a heavy measurement culture collect all kinds of metrics. The problem is that they may not necessarily connect or convey meaningful information in a coherent way.

Stakeholders and beneficiaries of the PMO can also be challenging, as they may not be sure about the real purpose of the PMO or what they want from the PMO. In this sense, they are also like typical customers who are guilty of changing their mind of what constitutes value to them. A chief information officer that we worked with complained that the portfolio management process that the PMO had been implementing over the last year, of which he was a big proponent, was not adding value. He was considering disbanding the process because it did not solve his problem of getting additional resources from the steering committee. He was no longer sure of the purpose of the PMO anymore. He was not willing to acknowledge the broader impact and the intangible value of better relationships with the business customers, among many things this PMO process was enabling.

PMO value is invisible, which makes the stakeholders forgetful of how things have changed and unlikely to attribute it to the PMO. Like water to the fish or air to humans, the PMO's value is transparent and often taken for granted.

The above factors highlight the challenges with measurement and why effective measurement remains elusive. Everyone tries to measure; some measure too much, some too little. What gets measured doesn't matter, or worse, drives the wrong behavior. To measure effectively you must decode, and work on the DNA strands of measurement.

THE DNA STRANDS OF MEASURE

Figure 8.1 lists the DNA strands of measure.

Objectives

To start with, you have to know how to define success. Objectives help to define success. Effective objectives depend on two aspects: questioning the clarity of purpose and perspective of measurement.

Purpose

Remember Karen from the beginning of the chapter, she was all excited about her agile metrics and burndown charts, and her manager kept asking, how many stories have we completed? You can see how there is nothing wrong with the measures themselves, but the measures did not help them to go live on time. The customer was not happy, and they could not bill and realize the revenue in this quarter. The question of how many stories were completed, limited Karen's view. Questions limit our frame and focus of what we notice and measure. If Karen's manager was clear on the definition of success and purpose, he should have asked, "How many stories can we release?" which assumes that the testing and quality assurance has been completed and the production environment is ready. A better question might be, "How many stories are we happy to release? Or, even better, "How happy and satisfied will the customer be with this release?" or "How much value will the customer experience with this release?" The quality of your question will help clarify the purpose and define the effectiveness of your measure.

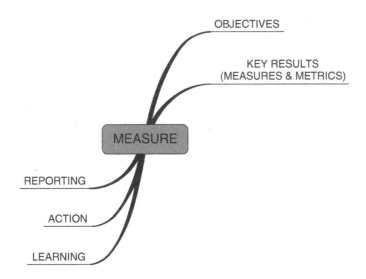

Figure 8.1: DNA Strands of Measure

In project management, there is a limited view of measurement—primarily to monitor, track and report status, and sometimes to forecast and predict the status of deliverables and outputs. The purpose of measurement is not just to measure but to provide feedback for improvement. If the purpose is clear, it will spark the crucial questions. Effective measurements should help:

- Influence behavior and drive performance toward desired outcomes and results
- Provide feedback to adjust and adapt
- Enhance customer satisfaction and experience
- Set program/project priorities
- Allocate resources effectively
- Gain insights
- Make better decisions and prioritization
- Provide transparency
- Communicate and inspire action
- Demonstrate benefits and value
- Illuminate impact

Perspective—Whose Perspective Are We Measuring From?

If the customer is the center of the universe and the purpose of business is to create customers, we must question whether our measures are customer- or stakeholder-centric (outside-in) or self-centric (inside-out). We tend to measure things that are important to us, like internal process efficiencies, versus customer success, or what matters to them. While you grapple with measures like customer satisfaction, customer success, and customer loyalty, which might all lead to a higher net promoter score, it all depends on what the customer perceives as value. That's why before selecting measures we must define and understand value.

Merriam-Webster's dictionary defines value as fair return or equivalent in goods, services, or money for something exchanged, **relative worth, utility, or importance**. We need to emphasize *relative worth, perceived value,* from the receiver's standpoint. The ultimate measure of the value is the sum of the benefits that accrue to each of the internal stakeholders from the stakeholder's perspective.

Even though we may know this definition and understand it intellectually, in practice the tendency is to focus on the first part and often forget *the stakeholder's perspective.* Efforts to show project and PMO value fall into this trap and present value *inside-out* from the PMOs perspective, instead of *outside-in* from the stakeholders' point of view. So when PMO leaders and team members are asked to show how they are providing value they enthusiastically rush to show and tell all the accomplishments from their own perspective, such as how many processes they have rolled out, how extensive their PM methodology is, or how many project reviews they have conducted, without considering if their stakeholder's care about any of these things.

Ken was proud of his PMO accomplishments and thought he had the measures to prove it. He was surprised to find that the executives didn't care and continued to question the PMO value. After we worked with him and helped the PMO implement some of the points made in this chapter, the light bulbs went on and they could see why they were struggling. The measures that they were reporting were PMO-centric and not customer or stakeholder-centric. Nobody outside the PMO cared about how many people complied with the methodology or how many people were trained, even though they knew there could be a correlation to project success. The PMO did not discuss and establish the definition of success or PMO value. It did not attempt to measure and illuminate business value.

Value, like beauty, *is in the eye of the beholder.* The PMO may think it is providing great utility, but do the stakeholders or the receivers think so? Usefulness must mean something to the receiver, not the provider. That is why in the pursuit to demonstrate value we must always start with the satisfaction of stakeholders, beneficiaries, and end users of PMO activities and services. Understanding satisfaction is tricky because it is based on expectations and perceptions, which can be fuzzy, shifting, and elusive. Ultimately, satisfaction depends on stakeholders' perception minus their expectations. Perceptions have to be higher than expectations for a greater degree of stakeholder satisfaction. If expectations are high and perception is low, that is obviously a problem. For example, if the PMO is sold as the panacea to all project-related issues and six months into the PMO implementation there is some progress, but projects are still having problems, PMO satisfaction is going to be low despite improvements in some areas. Expectations and perceptions need to be managed by identifying what is important to the stakeholders. Any efforts to show value have to start by stepping into the shoes of the stakeholders and feeling their pain and focusing on how that can be addressed to their satisfaction. Some of the stakeholders, beneficiaries, or customers of the PMO may include project managers, functional managers, end customers, senior management, business partners, vendors, and contractors, among others.

One of the themes emphasized in this book is customer experience, and one of the measures for customer experience management is Net Promoter Score®(Figure 8.2). For more information, refer to https://www.netpromoter.com/know/.

The combination of clear purpose and right perspective leads to a better definition of success with specific objectives.

Key Results

Objectives help to define success, and key results help to measure it. Key results focus on measurement. This strand addresses what to measure. What are the consequences of what we measure? How do we measure?

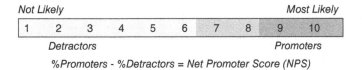

Figure 8.2: Net Promoter Score

The biggest challenge is deciphering what to measure. Finding the right measures and metrics and the related algorithm is the key to effective strategy execution. What you measure puts the spotlight on that factor, but how do you know if that is the right thing? What about the things you are not measuring? Measurements drive behavior, and every measurement has a consequence. To select effective measures and metrics, start by questioning: Do current measures describe relevant current and past status and performance data? Do current measures help predict and forecast future outcomes and results? What leading metrics will help drive future performance and results? How can we link and balance metrics from all elements of the DNA of strategy execution? Do current measures from other areas of the DNA cascade and align to strategy? Do we balance between output and outcome measures? What is the key measure and related algorithm that can have a core impact on performance and results? What is the overall impact of our measures?

Measure versus Metric

The terms *measure* and *metrics* are often used interchangeably. There is an overlap, but there is a difference. Measures are concrete and usually measure one thing—quantitatively, for example. We deployed five resources. Metrics describe a quality and require a measurement baseline; for example, we deployed 10 percent more resources on this project than the previous one. Measures are useful for demonstrating workloads and activity, and metrics are useful, particularly in the PMO, for evaluating compliance, process effectiveness, and measuring success against established objectives [https://cio .gov/performance-metrics-and-measures/].

"We tend to overvalue the things we can measure and undervalue the things we cannot."

John Hayes

Are You Ben or BoB?

What you measure depends on your measurement mindset or persona, whether you are like Ben or BoB? Let's meet Ben first. He is a by-the-book project manager, and when you ask him about the status of his project, he is ready with his project measures and will rattle off ... *"the cost performance index (CPI) is 0.96, and the schedule performance index is 1.06."* BoB is another project manager, and his response is different, *"Just met with the customer and the project is going to be delayed, but we uncovered a key customer functionality, and we can get it done, and this will have a positive impact on our revenue, about 15 percent better than we had projected."* Ben is focused on output measures; BoB is focused on outcomes. Ben is caught up in measuring process and activities, and Bob is targeting results. **Ben** stands for the **Benefit** and **BoB** stands for the **Benefit of the Benefit.**

Ben is limited to the benefit of the output, whereas BoB shows the real value related to strategic outcomes. For example, project measures confined to time, cost, scope or quality are Ben (output) measures. Whereas return on investment, time to market, or customer satisfaction measure what are Bob (outcome) oriented. Similarly, in the PMO realm, emphasizing the benefit of increased compliance, or increased number of trained, certified project managers, without showing how that translates into greater project success or strategic benefit for the organization is limiting and does not highlight business value. Other examples include measures such as the number of visitors or mobile app downloads that are Ben-oriented, versus real-value BoB-oriented metrics, like retention percentage, monetization, and net promoter score (NPS).

You can channel Ben and BoB by focusing on the distinction between outputs and outcomes. Outputs are the deliverables or products/services like PM methodology is a product or service of the PMO. Outcomes, on the other hand, are the success criteria or the measurable result of successful completion of the output. For example, 100 percent consistency in projects due to the standard use of the PM methodology. All too often, there is a great deal of emphasis on collecting outputs, and not much emphasis on outcomes. In fact, outputs by themselves may have no intrinsic value, unless they can be converted to outcomes. Outcomes can be further classified as benefits, which are the measurable desired results of the outcomes. One good test is for each measure to complete the sentence, "This _____ (output), will result in _____ (outcome)."

Ben and BoB mindsets apply to many things in life. Many people are limited with the Ben mindset and can't think beyond outputs and seem to be satisfied with it. The people and organizations with the BoB mindset go farther. Ben is satisfied with the near term, what's in view, within reach, easily measurable benefit, which is often limited.

Ben is worried about the form; BoB is after the essence. While Ben is busy measuring and analyzing the finger pointing to the moon, BoB is calculating what needs to be done to reach the moon. BoB's approach is simple and strategic—how to move the needle and widen the wedge (discussed in Chapter 4). He does that by focusing on three things: how we can keep or create more customers (increased revenue), save money (reduce costs), and not get in trouble (cost avoidance). BoB frames or translates everything Ben executes in one of these three aspects.

The question is, who do you want to be like, Ben or BoB? A quick response might be, "more like BoB," but as you reflect on the nuance, you realize you can't get to BoB without Ben. Ben is execution oriented, and BoB is strategy focused. Neither is better than the other; we need both. If you are more Ben oriented, you need to start to zoom out and challenge yourself to see the world through BoB's eyes. If you are BoB, you have to understand you cannot realize your strategy without Ben and develop some Ben capabilities or make sure you have some Ben's on your team. The ideal is BobbyBen, who is bimodal and balanced and can leverage both aspects to achieve desired results and make an impact on strategic execution.

Table 8.1 lists measures that will help to strike a balance between different dimensions of measures.

Table 8.1: Outputs versus Outcomes	
Outputs (Ben)	**Outcomes (BoB)**
Tangible (hard)	Intangible (soft)
Direct	Indirect
Short-term	Long-term
Quantitative	Qualitative
Objective	Subjective
Content-specific	Context-specific
Lagging	Leading
Operational	Strategic
Procedural	Behavioral

In developing a holistic approach to measuring and showing value, the challenge is to be able to include the opposing dynamics while striving to strike a balance between different types of measures. Often, only one side of these factors is taken into consideration. Many perspectives should be considered in developing a balanced approach to measurement.

The matrix in Figure 8.3 illustrates classification and examples of different types of measures and metrics and the need to balance.

Another aspect that should be considered in selecting metrics is the importance of the end-to-end result instead of a small part of the process. In their book, *Implementing Lean Software Development*, Mary and Tom Poppendieck describe this as optimizing the whole, which means ensuring the metrics in use do not drive suboptimal behavior toward the real goal of delivering useful software.

Objective   Subjective

	Objective	Subjective
Quantitative	10 story points completed in this sprint	10% increase in customer satisfaction
Qualitative	Portfolio committee voted & approved 10 projects based on business case review	Customer likes early delivery

Figure 8.3: Different Types of Measures

Decoding the Algorithm that Leads to Desired Results: Focusing on What Influences and Drives Ben and BoB

It is important to balance Ben and BoB measures, but you have to hope or pray that you achieve them. Ben measures like on-time or under-budget delivery, or BoB measures like revenue, margin, and profitability are the goals or objectives of what to achieve. You can set the target, but you don't know if you will achieve it, and by the time you measure and find out it is already late, as these are in the past or lagging measures. This is the difference between lag and lead measures. You should also focus on how what you measure can influence Ben and BoB and what measures will drive their behavior toward achieving the objective. The key is to identify the leading measures that influence the achievement of Ben and BoB. You must ensure not just that you measure the right thing but that your measures are influencing the right behaviors.

Lag indicators are easy to measure but hard to improve or influence, while leading indicators are typically hard to measure and easy to influence. Lead indicators are more difficult to determine than lag indicators. A good test is to ask, does the measure allow you to influence the result? This is also the secret to decoding the measure and related algorithm that can lead to desired results.

The challenge is to identify the key measure that will influence the right behavior; for example, in running, instead of trying to focus on the goal of a seven-minute mile, measuring cadence—steps per minute. Or in basketball, instead of focusing on making 10 shots in a row, focus on keeping the elbows in. If you are trying to lose weight, instead of focusing on number of pounds, focus on number of calories consumed per day or number of workouts per week. The key is to make sure it is a specific measure that is predictive. For example, instead of *improve concentration,* the measure should be *number of passes tossed correctly.*

Finding the right measures and metrics decoding the algorithm that leads to success is not easy but should be the top priority of executives and managers at every level.

Once the right measure is identified, create a key measure profile for each measure as illustrated in Figure 8.4 to bring more rigor and credibility to the process.

Measurement rigor adds overhead, which needs to be balanced with the cost of precision, accuracy, robustness, and the value of the metric.

Once you know what to measure, the next challenge is, how do we measure? Do you have the right systems, tools, apps, or sensors to collect the data? We can get all enamored with the colorful dials, dashboards, and scorecards of measurement systems, including project portfolio management (PPM) systems. The challenge is that they are still dependent on humans to enter the data to provide meaningful information. These systems can't capture what is truly going on because of missing, incorrect, or nuanced information. In a world of wearables and sensors, the tools and apps are getting better at collecting and tracking data, but the PMO has still a lot of work to do to ensure the timely input and quality of data.

Measure	Effective Date	Required PM Framework Project Completion Criteria Updated	7/1/0XX
Measure Owner			
Business Objective			
Metric (Supported Key Success Factor)		Consistent utilization of PM Framework (PMF) Methodology	
Measure Definition		Percentage of projects with completed PMF criteria	
Frequency of Measurement		Monthly	
Unit of Measure		%	
Calculation		Total number of projects with completed criteria divided by Total number of projects sampled multiplied by 100	
Polarity		Higher is better	
Data Sources		■ PMF Logs ■ Project Reports ■ PMO Reports ■ Project Documentation	
Data Collection Process		Randomly select specified documentation from 50% of active projects and perform measure calculation	
Data Collector			
Performance Baseline		TBD; Collect baseline data using results from July, August, and September samples.	
Performance Target		100%	
Performance Enablers		Utilizing most recent version of deliverables/templates and continuously working the documents on a monthly basis, education and documented process	

Figure 8.4: Key Measure Profile

Reporting: Presentation and Communication

How Do We Present and Communicate the Measures and Metrics to Influence Desired Action?

To make sense and mean something, measures are presented and communicated in reports, dashboards, and scorecards. They tend to show current values of a few metrics taken out of context with little or no history. While well-meaning, they are often misinterpreted and confusing. It is hard to decipher the signal from the noise. The renowned guru of information design, Edward Tufte, explains, "Clutter and confusion are failures of design, not attributes of information. There is no such thing as information overload." He advises, "The purpose of analytical displays of evidence is to assist thinking. Consequently, in constructing displays of evidence, the first question is, what are the thinking tasks that these displays are supposed to serve?"

Traffic Lights

A common staple of communicating status in the project world is traffic light dashboards. Do you use a traffic light approach to communicate project status? Do you get frustrated at times that the color does not necessarily represent the true status of your project? Do you spend a lot of time defending the color, and find that even after detailed explanation the status is misunderstood by your project stakeholders? You are not alone.

Traffic light project status reports are popular with senior managers and executives, but they are also a source of misinterpretation, confusion, and frustration. Red, amber/yellow, and green—also known as RAG—reports are widely used because they are a simple, visual way to communicate project status. Just like traffic lights on the street are wired to a timing mechanism that causes the light to change color, the status reports should be wired to objective criteria. For example, if a project is within a 5 percent variance, it may be green; between 5 percent and 10 percent, yellow; and greater than 10 percent, red.

Unlike street lights, though, project status lights are often misinterpreted. Typically, nobody likes red status on a project, taken out of context; it is interpreted as bad, so everybody wants to see green. But contrary to conventional thinking, red may not necessarily be bad and green may not necessarily be good for your project. Red may mean that there are some changes or issues caused by customers or stakeholders that have pushed the project past the variance threshold. Despite this, it may be good for the overall achievement of the project objectives. It may mean that the project manager is doing a good job of engaging and listening to the stakeholders, or that the project team is not complacent.

On the other hand, green may be a symptom that the project team is narrowly focused and is underachieving the project goals. It can also signal that the project is being re-baselined frequently. A prolonged green status can also make the project team too comfortable, and not prepared for unforeseen risks and issues.

Because of these misconceptions, traffic light status reports are prone to be gamed by project managers, who learn to manipulate the colors to fit their management's expectation and organizational culture. There can also be mistaken expectations, since, while traffic lights change in a linear, predictable pattern, projects can change state at random, and jump from green to red in an unpredictable way, often without warning.

It may be a good idea, then, to try a different approach, even if you find it difficult to abandon the traffic light reports because senior managers are used to them. Instead of traffic light colors, use rich data and numbers with sparklines (a very small line chart, typically drawn without axes or coordinates, made popular by Edward Tufte) and patterns to communicate status.

If you are compelled to use traffic lights, here are some tips to make your status reports more effective:

- Calculate the overall health of the project with a balance between objective measures like schedule variance and cost variance, and subjective measures like number of issues and stakeholder engagement.
- Periodically review and recalibrate the variance thresholds that trigger project status change.
- Provide a balanced perspective by using two lights—one based on the objective calculation of defined thresholds, and the other a subjective light based on what the project manager feels is the true status of the project.

Context, Contrast, and Causality

Practice and clarify the 3 C's—context, contrast, and causality—in your presentation medium to raise the communication value of your measurement. As you saw in the traffic light example, the color without context can be misinterpreted. To provide greater context, use comparison and contrast. For example, this project is over budget, but compared to project Y, this is 5 percent less, or compared to last year we are 7 percent ahead. Misattributing causality and linkage is a common flaw in communicating and interpreting metrics. Clarify and communicate, and emphasize and underscore whether there is linkage or not. Recent project delays are not linked to the new PMO tool or cost savings due to new governance process.

Information Radiators

Agile uses the term *information radiator* (coined by Alistair Cockburn) as a generic term to promote transparency of information and reporting in an agile environment. It can include any number of handwritten, drawn, printed, or electronic displays in a highly visible location so that all team members and stakeholders can see or access the latest information at a glance: burndown charts, count of automated tests, velocity, incident reports, continuous integration status, and so on. Information radiators promote a sense of responsibility and ownership among the team members. Based on the premise that the team has nothing to hide from itself, or customers and stakeholders, information radiators tend to provoke conversation, to openly confront and solve problems.

Action—What Can We Do with the Measures? Are the Measures Actionable?

While there are myriad measures and metrics you can collect, track and analyze, but the only ones that are consequential are the ones that inspire action. Often, measures have the opposite effect, as we are numbed by the numbers and don't know what to do. Although the various dials and meters look interesting in the report, they are not actionable. Measures are useful only if they are used for actionable treatment, rather than just for autopsy.

Here's an experiment you can try to find out if your metrics are actionable or not. If you or your PMO is responsible for preparing and publishing a bunch of reports, pick a couple and stop sending them. If nobody complains, you know they were inconsequential and not needed.

To be actionable, it must be a leading metric that has a direct impact on performance and can be influenced. For example, for one software development team, the number of stories accepted, became a key measure that had an overall impact. They changed their behavior to have more interactions with the customer, uncover hidden needs, and focus on acceptance, not just completion of story points.

Learning—Are We Getting Effective Feedback? Are We Learning and Adjusting?

The measure that keeps Kumar (from one of the opening scenarios in this chapter) up at night is the number of bugs fixed. It seems like a never-ending cycle as he fixes a few, and finds more added to the list. How can Kumar get out of this downward spiral? The number of bugs measured is trolling him in a repeated loop, without questioning what the goal is and why the bugs are occurring in the first place? He does the something over and over with bug fixes with little hope of getting out of the loop. The measure is not implemented intelligently with a complementary measure of reduction in number of bugs, rather than bug fixes alone. To sharpen measure intelligence Kumar needs to analyze the measure and the underlying forces. He can apply **double-loop learning,** which entails the modification of goals or decision-making rules in the light of experience. The first loop uses the goals or decision-making rules, and the second loop enables their modification—hence "double-loop." According to Chris Argyris, Professor Emeritus, Harvard Business School, double-loop learning recognizes that the way a problem is defined and solved can be a source of the problem

The purpose of measurement is not just to measure but to learn and improve. Improve both results and measures themselves to achieve the desired impact. That means understanding when measures are no longer relevant, replacing them, or dropping them as progress is made toward the goal, or the environment has changed. Chapter 10 delves deeper into different aspects of learning and feedback.

DEFINING AND MEASURING PROJECT SUCCESS

The traditional measure that project managers are expected to manage is the triple constraint. They are often compelled to live in this triangle of time, cost, scope, and quality. The initial idea of the triple constraint was a framework for project managers to evaluate and balance these competing demands. It became a way to track and monitor projects. Over time, it has also become a *de facto* method to define and measure project success. While the triple constraint is necessary, it is not enough. Projects that are delivered on time, within budget, and meet scope specifications may not necessarily perceived to be successful by key stakeholders.

Besides time, cost, scope, and quality, what are other criteria for project success in your organization? We have surveyed this question of various project stakeholders over the past 10 years and repeatedly heard that ultimately the factors contributing to project success include:

- Stakeholder and customer satisfaction
- Meeting business case objectives
- Time to value
- Customer/end-user adoption
- Customer experience
- Quality
- Meeting governance criteria
- Benefits realization
- Overall value and impact

As you already recognized, time, cost, scope, and quality are related to project outputs (Ben), whereas these other factors are related to business outcomes (BoB). While the triple constraint is important, it can also narrow the focus away from other crucial factors that lead to project success. Based on today's project environments, project managers need to broaden their perspective to include other criteria to satisfy stakeholders and deliver business results.

How do we rethink the triple constraint? We have posed this question and conducted a number of exercises on this topic. Following are three overlapping perspectives that should be reviewed and discussed as you come up with ways to define and measure success:

1. **Mirror project outputs with business outcomes (Figure 8.5).** While focusing on each of the triple constraints, the project manager has to reflect and make project decisions based on the achievement of the corresponding business outcome. Cost and time focus must optimize business benefits like return on investment (ROI), net present value (NPV), and so on, and benefits of faster delivery or time-to-market. The scope should mirror end-user adoption, and overall quality must be balanced with stakeholder/customer satisfaction.

2. **Parallel balance (Figure 8.6).** This is a parallelogram view of balancing between scope and schedule in parallel with budget and benefits, or budget and scope in parallel with schedule and benefits. Benefits may include a combination of business objectives, end-user adoption, customer satisfaction, and other criteria.

3. **From a triangle of constraints to a diamond of opportunity (Figure 8.7).** The diamond combines the delivery focus of project outputs on one side, with business outcomes on the other. The idea of the multiple sides of the diamond also helps the project manager to include multiple perspectives of focus that might be relevant to their business. You can select a combination of factors to measure that may be relevant to your business and industry. For example, in many organizations, an additional criterion that is becoming a critical balancing factor similar to quality is health, safety, security, and environment (HSSE).

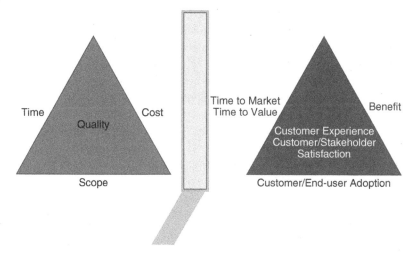

Figure 8.5: Mirroring Project Outputs and Outcomes

Parallel Balance

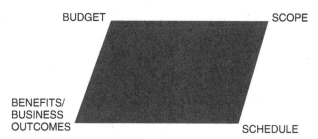

Figure 8.6: Parallel Balance

Figure 8.7: Diamond

If you are expected to focus on the triple constraint alone, expanding your focus to include additional criteria will give you an edge. You will be able to deliver projects that provide business benefits, and your projects will be perceived to be successful by your stakeholders.

What about agile? It is a little better, but agile metrics like burndown charts and velocity are also primarily output oriented. Agile turns the triple constraint upside down. The scope is not fixed at the start. You work with fixed time (iterations) and cost (resources) and adjust the scope. The goal is to prioritize and work on the customer's most important requirements within the budgeted cost and time. The good thing is that the customer is involved, and prioritization is based on business value. But you don't know if the customer will be satisfied or use it, and you don't know if you will achieve the business outcome. Unless you broaden the perspective and include leading outcome measures like adoption and experience that lead to business outcomes.

If you are thinking, "Wait, that is not the job of a project manager; projects deliver outputs and programs deliver benefits," think again. This is a misconception, as we discussed in the beginning of this book, outputs and outcomes are both products of the same DNA of projects, programs, and portfolio. If anything, they should be identified and linked together, the breakdown and separation are what cause the mirror to crack further, and it is hard to put it back together. Many organizations even have a hard time distinguishing between projects and program. If the organization does not define program, does that mean it's projects cannot deliver any benefits?

Whether it is projects, programs, strategic initiatives, portfolio, or PMO, there should be an integrative approach to measurement, and you must start by defining success with your customers and stakeholders.

OBJECTIVES AND KEY RESULTS (OKRs)

Objectives and key results strands of measure are derived from the OKR technique, which is based on Peter Drucker's management by objectives (MBO) framework, repurposed, and tweaked to twenty-first-century organizational dynamics. First implemented at Intel by Andy Grove in the 1970s and later adopted by Google and other companies like LinkedIn, Oracle, Twitter, and many other Silicon Valley companies, OKRs track individual, team, and company goals and outcomes in an open and transparent way.

Objectives define the purpose and where you want to go in a set time frame. They are qualitative and describe the desired outcome. They should inspire action, so they should be inspirational with a stretch goal and actionable, which a person or team can execute. Key results add metrics to objectives, they are quantitative. They let you know if you are getting there, or how far you are from your goal.

OKRs are about focus, frequent feedback and response, stretch goals, transparency, and simplicity. As a result, insights, and improvements are easier to see and implement.

OKRs promote a sense of ownership, as they are determined bottom-up as opposed to key performance indicators (KPIs), which are top-down. OKRs should be transparent so that everyone sees the bigger picture and can hold each other accountable.

Implementing OKRs:

1. List three to four objectives you want to strive for.
2. For each objective, list three to five key results to be achieved.
3. Communicate objectives and key results to everyone. They should be easily accessible and transparent.
4. Regularly update each result on a 0 to 100 percent scale.
5. When objective's results reach 70 to 80 percent, consider it done. They are stretch goals and should be designed to challenge.
6. Review OKRs regularly and set new ones. They should be set quarterly and reviewed monthly.

Example:

Objective: Improve PMO value and impact by the end of Q1

Key Result: Increased interaction with key customers and stakeholders—conduct survey with 85 percent or better response and 15 one-on-one interviews with key stakeholders

Key Result: Three enhancements and two new PMO processes implemented

Key Result: 90 percent adoption of PMO standards

Key Result: 7 percent project cost reduction due to greater standardization and consistency

DEVELOPING AN INTEGRATIVE APPROACH: STRATEGY EXECUTION MEASUREMENT FRAMEWORK

An effective measurement framework should focus on multiple dimensions and optimize the whole. For example, how do you know if you are covering all the critical areas when developing OKRs? You may unintentionally isolate some or overdose on others. Measurement should cover and balance all aspects of the DNA—Strategy, Execution, Governance, Connect, Measure, Change, and Learn. Also, how do you know if you are heavy on output (Ben) measures, versus outcome (BoB) measures, or how do you fine-tune what lead measures influence Ben and what influences BoB.

The Strategy Execution Measurement Framework provides an integrative approach to develop metrics and enhances the practice and application of OKRs with a holistic perspective. It promotes a sense of ownership and linkage of results to outputs, outcomes, and impact. It is designed to ensure that you measure both, outputs and outcomes, and the leading measure to achieve them.

Start by collaborating with customers and stakeholders to define success and come up with OKRs. Whether it is a strategic initiative, program, project, or the PMO, define meaningful metrics by listing the organizational business objectives in the background, and follow these steps. Define:

1.0 Business/Organizational objective

1.1 Objective that needs to be achieved (what is being developed or worked on)

2.1 Key result—outcome measure or metric for what you are trying to achieve (BoB) [Outcome Measure]

2.2 Key result—lead measure or metric that will lead to desired outcome (BoB) [Lead Outcome Measure]

2.3 Key result—output measure or metric for the deliverable or outputs you need to achieve the outcome (Ben) [Output Measure]

2.4 Key result—lead measure or metric that will lead to desired output (Ben) [Lead Output Measure]

3.0 Customer/Stakeholder impacted (who cares about this)

This approach is designed to make sure you breakdown key results into outcomes (BoB) and outputs (Ben) and what leads to both, without missing any one or overdosing on the other. The order in how you get to them might seem confusing, but once you start to think it through, you will find an approach that works best for you.

Here's an example.

Let's say the PMO has identified a pain point with no standardized project management approach. There is an existing PM process, but it is not effective. Along with the stakeholders, the PMO comes with an objective to implement an effective PM framework that has an impact within three months.

Typically, PMOs would mostly measure how many people are trained and certified in the new framework (Ben). They might be trained and certified, but that does not necessarily mean that they will use and apply the process. They might even apply the process but might not be aware of the purpose or outcome the PMO is trying to target (e.g., key result—outcome of 10 percent cost reduction; BoB).

The above approach helps to start with the question, why implement the PM framework? To achieve the key result of overall cost reduction by 10 percent due to increased standardization, consistency, and less wastage (outcome measure). To realize this, you have to measure the adoption and application (80 percent) or above of the new PM framework (lead outcome measure). This will not be possible unless you measure how many project managers are certified (85 percent) in the new framework (output measure). What will lead to certification is measuring the number of project managers (95 percent) trained in the framework (lead output measure).

Here are the steps:

1.0 Business/Organizational objective—10 percent cost reduction

1.1 Objective (PMO)—effective PM framework in three months

2.1 Key result—outcome measure (BoB)—10 percent project cost reduction due to increased standardization, consistency, and less wastage

2.2 Key result—lead outcome measure—80 percent adoption and application of PM framework templates

2.3 Key result—output measure (Ben)—85 percent compliance with PM framework

2.4 Key result—output lead measure—95 percent project managers and team members trained and certified in PM framework

3.0 Customer/Stakeholder impacted—project managers, team members, sponsors, project customers, PMO

Different stakeholder may care about different measures, and focus on measures that they can target and improve.

To better understand the steps, Figure 8.8 provides a simple example. Imagine you just had a physical and you were diagnosed with hypertension or high blood pressure (BP), and you have to get in shape. Instead of arbitrarily thinking about exercising, or diet, how could you use better measures and OKRs to get in shape? You should discuss the objective with your doctor of achieving better health and a normalized blood pressure in three months. The key result you want to target is normal BP (BoB—outcome measure). What is going to lead to it? You have to make sure you measure and track your BP regularly (lead outcome measure). That is fine, but what will contribute to a lower BP? Losing weight by 20 pounds will help me get in shape and also impact my BP (Ben—output measure), and what will lead to losing weight? Calorie intake and/or walking and tracking number of steps per day (lead—output measure).

As you measure and track you will know if your walking is impacting your BP. It will also help you to regulate how many steps I need to walk to achieve the desired weight. In some cases, weight loss might not be effective; managing diet and sodium intake might be more effective. Some measures can act at cross-purpose and drive undesirable outcomes. Once you identify the Ben and BoB measures, the challenge is not to stop there but also determine the right lead measures that will lead to them or the desired outputs and outcomes.

Figure 8.9 is a template that can be adapted for projects, programs, portfolio, or PMO. Additional columns like Progress, Actual Results, or Assumptions can be added to the template to create a scorecard.

1. Objective	2.1 Key Result Outcome Measure (BoB)	2.2 Key Result Lead Outcome Measure (Leads to BoB)	2.3 Key Result Output Measure (Ben)	2.4 Key Result Lead Output Measure (Leads to Ben)
Better health (normal blood pressure) by focusing on weight loss in 3 months	Normalized BP	Daily monitoring & tracking BP	Lose 20 lbs.	Walk 10,000 steps per day

Figure 8.8: Better Health Measurement Example

DNA Element	Objective		Key Results				Customer / Stakeholder
	1.0 Business / Organizational Objective	1.1 PMO or Program or Project Objective	2.1 Key Result Outcome Measure (BoB)	2.2 Key Result Lead Outcome Measure (Leads to BoB)	2.3 Key Result Output/ Deliverable Measure (Ben)	2.4 Key Result Lead Output Measure (Leads to Ben)	3.0 Who Is Impacted (Who cares?)
Strategy (How will we know if we are in alignment & achievement of org. strategic objective?)							
Execution (How will we know if we are executing effectively?)							
Governance (How will we know if we are meeting gov./compliance criteria?)							
Connect (How will we know if we are connecting & engaging key customers & stakeholders?)							
Measure (How will we know if we have effective measurements & reporting?)							
Change (How will we know if we have effective change management in place?)							
Learn (How will we know if we are effectively learning?)							

Figure 8.9   Strategy Execution Measurement Framework
Source: © J. Duggal. Projectize Group.

Figure 8.10 provides illustrative examples for PMO measures and how to use this template for PMOs. To start with, list the business/organizational objective (1.0), collaborate with stakeholders to define success with PMO objectives (1.1) in the DNA areas. It is important to start with the areas with the most pain. Next, discuss and list the key results you want to achieve in a given time frame. Next, classify the key results and make sure you are covering outputs or Ben (2.3), and outcomes or BoB (2.1). Then also make sure you discuss and list what will lead to BoB (2.2), and what will lead to Ben (2.4). In the beginning, it can seem confusing, but as you use it, you will find your own way that works for you. The idea is that the template simply provides a way to focus on outputs and outcomes and what leads to them. Also, it is a visual tool that highlights the areas that you are working on, and whether you have balanced output and outcome measures, and what leads to them.

Figure 8.11 provides an example of how to use this framework for a project. As we discussed earlier, typically project measures primarily focus on one element of the DNA—Execution—and only on output measures, like on-time, within budget execution. This framework provides a visual template for a balanced approach covering outputs and outcomes, with lagging and leading measures, with a broader perspective including the other elements of the DNA, which are often not measured and thus not managed as well. To apply, collaborate with your project sponsor and key stakeholders to define project success objectives for the DNA areas. Start with execution and strategy, and then you can add other areas as appropriate. It is not necessary to define all of the areas and each of the key results columns in the beginning. The sample measures in the figure are only illustrative. Your specific measures will depend on your context and what your stakeholders care about. As you progress, the template provides a visual means of highlighting your focus areas and measurement.

In some instances, it may be hard to quantify outcomes and link them to impact on revenue or cost savings, but it is still effective to have a target. In these instances, focus on the outcome lead measure that will ultimately impact revenue or cost. No doubt, this is a tough exercise, but it is worthwhile because it forces the linkage and measurement of all elements of the DNA in a holistic way. To start with, you do not have to complete OKRs for all elements of the DNA. Focus on a couple of the pain-point areas and add others as you evolve. Also, measures for some elements may not be relevant in every instance.

MEASURING AND SHOWING PMO VALUE

Showing PMO value continues to be a top PMO challenge. 39 percent questioned the existence and need for a PMO, according to an ongoing PMO study by the Projectize Group. The belief that PMOs improve overall project management effectiveness and contribute to project success may be true. But it is challenging to show it. PMO value is elusive at best, as it is often hidden in improvements and efficiencies that are intangible

DNA Element	Objective		Key Results				Customer / Stakeholder
	1.0 Business / Organizational Objective	1.1 PMO Objective	2.1 Key Result Outcome Measure (BoB)	2.2 Key Result - Lead Outcome Measure (Leads to BoB)	2.3 Key Result Output/ Deliverable Measure (Ben)	2.4 Key Result - Lead Output Measure (Leads to Ben)	3.0 Who Is Impacted (Who cares?)
Strategy (How will we know if we are in alignment & achievement of strategic objective?)	Cost-reduction by 10%	Alignment of Projects to strategic objectives – Review and improve portfolio selection & prioritization process	Cost savings of X% due to discontinuation of nonaligned projects	Monthly portfolio review & prioritization – targeted decision-making for X% reduction	X% projects meet strategic alignment criteria	X% of projects in portfolio pipeline process	Executives PMO Finance Project/Program Managers Other functional areas impacted
Execution (How will we know if we are executing effectively?)		Improve Project Success rate – Implement PM Framework by end of Q1	X% Impact on revenue 4% Cost savings (due to penalty cost avoidance)	X% Less rework Zero penalty cost	X % of projects meeting customer requirements	X% of projects applying std. PM framework	Customers Project / Program Managers Team Members Executives Finance PMO
Governance (How will we know if we are meeting gov. / compliance criteria?)		Ensure benefit realization of projects – Implement stage-gate process	X% impact on revenue due to focused benefits realization X% Cost savings	X% projects assessed for benefits realization stage-gate criteria & decision making to discontinue projects that do not meet benefits realization criteria	X% project compliance with stage-gate process	X% project & program managers trained in benefits realization & updated stage-gate process	Customers Project / Program Managers Team Members Executives Finance PMO Governance committee

Figure 8.10   Strategy Execution PMO Scorecard

Source: © J. Duggal. Projectize Group.

DNA Element	Objective		Key Results				Customer / Stakeholder
	1.0 Business / Organizational Objective	1.1 PMO Objective	2.1 Key Result Outcome Measure (BoB)	2.2 Key Result Lead Outcome Measure (Leads to BoB)	2.3 Key Result Output/ Deliverable Measure (Ben)	2.4 Key Result Lead Output Measure (Leads to Ben)	3.0 Who Is Impacted (Who cares?)
Connect (How will we know if we are connecting & engaging key customers & stakeholders?)		Improve PMO Stakeholder Engagement & Satisfaction	X% cost savings due to greater adoption of PMO cost-focused processes	X% increase in PMO Net Promoter Score (NPS) leading to wider stakeholder engagement	X% Stakeholder satisfaction ratings of higher than 4 on a scale of 1–5	Interviews with 90% of key stakeholders. Stakeholder survey with 85% response rate	PMO Stakeholders
Measure (How will we know if we have effective measurements & reporting?)		Measure PMO Value – Conduct PMO Delight Index (PDI) Survey [PDI explained in Chapter 12]	X% cost savings due to higher PDI & adoption of PMO	PMO Delight Index (PDI) above 3.5	PDI survey with X% response rate	X no. of PDI awareness sessions X no. of targeted social media X no. of incentives for survey response	Executives Functional Areas PMO Stakeholders
Change (How will we know if we have effective change management in place?)		Increase adoption rate of project deliverables / systems – Implement change management processes & training	X% cost savings	X% usage & adoption rate of deliverables 15% increase in customer satisfaction & NPS	X% projects/ programs apply change readiness assessment	Train X% of project/program managers and project teams in change mgt.	Project/Program Managers Team Members End users Customers PMO
Learn (How will we know if we are effectively learning?)		Continuous improvement by harnessing lessons learned – Implement collaboration tool	X% Cost savings due to greater reuse / cross-sharing of cost saving ideas	X no. of application of lessons learned stories Tool NPS above 4.5	X% adoption of collaboration tool	X no. of Collaboration tool awareness & training sessions attended by X% of project teams	Project/Program Managers Team Members PMO

Figure 8.10 (Continued)

DNA Element	Objective		Key Results			
	1.0 Business/ Organizational Objective	1.1 Project Objective	2.1 Key Result Outcome Measure (BoB)	2.2 Key Result Lead Outcome Measure (Leads to BoB)	2.3 Key Result Output/Deliverable Measure (Ben)	2.4 Key Result Lead Output Measure (Leads to Ben)
Strategy (How will we know if we are in alignment & achievement of strategic objective?)		Ensure alignment – Project deliverable aligns to achievement of business objective (X% impact on revenue)	Impact on revenue (XX Revenue target from project)	Track & measure alignment (X% Deliverables targeted for revenue goals)	Deliverables aligned with business case goals (X% Deliverables aligned with business case)	Review business case alignment goals (X% Review of business rev. completed)
Execution (How will we know if we are executing effectively?)	XX Revenue target	Successful project execution (within X% variance & X% impact on revenue)	Impact on revenue (XX Revenue target from project)	Customer & Product Owner Engagement (Meet X% customer requirements)	On-time within budget (within X% variance)	% complete Schedule Performance Index (SPI) Cost Performance Index (CPI) Velocity Burndown (within X% variance)
Governance (How will we know if we are meeting gov. / compliance criteria?)		Meeting required governance / compliance criteria	Cost avoidance No additional penalty costs	Meet X% risk assessment criteria	Pass project review/audit (4 or greater rating on review)	Compliance with governance reqs. (X% Completion of compliance checklist)
Connect (How will we know if we are connecting & engaging key customers & stakeholders?)		Improved stakeholder interactions (NPS score; No. of new customer acquisitions)	X no. of new customer acquisition	X% Customers at NPS score over 4.5	Meet stakeholder needs (Implement X no. of stakeholder requested improvements)	Meet weekly stakeholder meeting targets Meet stakeholder survey targets

Figure 8.11 Strategy Execution Project Scorecard

Source: © J. Duggal. Projectize Group.

DNA Element	Objective		Key Results			
	1.0 Business/ Organizational Objective	1.1 Project Objective	2.1 Key Result Outcome Measure (BoB)	2.2 Key Result Lead Outcome Measure (Leads to BoB)	2.3 Key Result Output/Deliverable Measure (Ben)	2.4 Key Result Lead Output Measure (Leads to Ben)
Measure (How will we know if we have effective measurements & reporting?)		Effective Project Reporting (timely availability of reports for effective decision making & course correction)	X Impact on revenue or cost savings due to timely course correction	No. of course correction actions taken based on timely data & reports	X% timely project reporting	X% timely data input
Change (How will we know if we have effective change management in place?)		Greater adoption of project deliverable	Impact on revenue (XX Revenue target from project)	X% Customer sat NPS score over 4.5	Increased adoption target by X%	Change readiness assessment & planning (Meet X% change readiness target)
Learn (How will we know if we are effectively learning?)		Effective learning and application of lessons learned	Cost saving due to application of lessons	Apply & improve process (X no. of lessons applied stories)	No. of lessons documented (Target no. of lessons input in collaboration tool)	Retrospectives (Meet X% retrospective target)

Figure 8.11 (Continued)

and hard to measure. Only 15 percent of PMOs even measure themselves in the survey mentioned above. Typically, reactive approaches are patched together to demonstrate the elusive benefits when the PMO is under the gun. There is a lack of a consistent approach to better substantiate the value of a PMO.

In my initial experience leading a PMO, I remember the frustrating part about measuring the PMO was that we knew we were providing a lot of good services—we could feel the value, but it was hard to show it. Like a typical PMO, we were improving project management processes, increasing efficiencies in project delivery, and providing project support and guidance, but it was always a struggle to identify and measure these benefits. A lot of the PMO's value is buried. The question is, how do we find the hidden value? Often, it is indirect and hidden in improvements and efficiencies that are hard to capture. There is a gap between the perceived value versus measurable value; 80 percent of respondents confirmed that their PMO had some value, but only 35 percent were able to measure it, according to our survey.

When we ask PMO teams, "What is the value you are providing?" they proudly list:

- Consistent methodology and standardization
- Repeatable processes
- Talent management trained and certified PMs
- Standardized tools and templates
- Projects on time and within budget
- Improved execution and delivery
- Lessons learned repositories
- Organizational change management
- Facilitating strategic initiatives

You can see how these are Ben-oriented. The challenge is to translate these into BoB-oriented results that executives, customers, and other stakeholders care about. How do you show the linkage of PMO activities and outputs to value provided? How do you show how the benefit of these activities will lead to increased revenue, cost savings, or cost avoidance and overall desired results and impact?

This chapter has provided a comprehensive approach to developing a measurement approach that can be used to identify and show PMO value. Following are some additional consideration:

Clarify the purpose of the PMO. The purpose of the PMO should be based on the overall context, business objectives, and strategy of the organization. If everybody is not on the same page regarding the purpose of the PMO, it may lead to measuring the wrong things and focusing on initiatives that do not align to the purpose of the PMO. Collaborate with your stakeholders to define success and value.

Identify value and selecting appropriate metrics to measure it. What is important to the stakeholders from their perspective needs to be identified and defined. The PMO needs to determine how it can address stakeholder needs and provide value to them. Metrics of how that value could be measured needs to be discussed and selected collaboratively with the stakeholders. Different sets of metrics may be required for different stakeholders. The current state needs to be baselined to track future changes and improvements.

Link the measures. It is crucial to show how the measures and metrics link to business objectives and related strategies. Additionally, any linkages to other stakeholder groups need to be investigated. Establishing links to the purpose of the PMO, business objectives, organizational context, business fit, and stakeholder needs is the key to individualizing and personalizing value from their perspective. The above measurement template links these elements in a simple way.

Illuminating PMO Value

PMO value is not obvious or is not going to surface automatically; it has to be illuminated and highlighted. As in other endeavors, to prove that we have accomplished a feat or have been somewhere, we take pictures. Similarly, the PMO has to develop and show snapshots of its accomplishments and the value it is providing from different perspectives. To articulate value, it must be packaged, marketed, sold, and communicated with a PMO marcom (marketing and communications) plan. Managing value should be a proactive process that can form the basis for managing stakeholders and their expectations. The PMO should not shy away from the value question; rather, the PMO should be actively engaged in creating a dialogue with the stakeholders by finding out what is important to them, explaining why some metrics are being collected and what you are going to do with them, and seeking ongoing feedback. Metrics should be used as a medium, not just a technique, for articulating and communicating value.

Chapter 12 provides a template for the PMO delight index (PDI) to measure PMO value.

Sustaining PMO Value: Managing Value Dissipation and Benefits Erosion

One of our client organizations was in the process of implementing a PMO for over two years, and they had successfully implemented several PMO initiatives, including portfolio management and a standardized project management methodology framework with a high degree of executive support and sponsorship. However, when a couple of projects were not on track, the PMO was under the gun, and the overall value of the PMO was questioned. PMO value can quickly dissipate if projects success rate does not improve and projects start to slip. There is a belief in the eyes of management that project execution should be successful and there should no longer be any issues with projects because the organization has a PMO. Management attributes any project failure to the PMO, regardless of whether PMO is directly responsible. Therefore, it is crucial for the PMO to set the right expectation regarding its role in successful project delivery; otherwise, it might erode the benefits and undermine perceived value in other areas as well.

Value depends on the credibility of the measurements. It can dissipate if there is suspicion around how that value is being calculated and reported. To establish tangible value, the benefits must be based on a sound premise and accepted assumptions. The cause and effect must be established based on a credible formula validated with proof and evidence. Refer to the key measure profile template, which helps define these elements and brings rigor and credibility to the process. If a governance policy to deal with intangible measures is in place, it can help in removing doubts about using intangibles. This governance policy can spell out whether intangibles should be considered, which kinds of intangibles, and how they can be used.

The perception of complexity or bureaucracy of PMO processes and templates and tools can lead to suboptimal usage, which erodes the benefits and diminishes value. Another reason why value tends to dissipate over time is the law of diminishing returns. As the benefits become transparent and integrated, they are taken for granted, and people change their perception and forget how bad things were without the PMO.

NEXT-GENERATION MEASURES

The move from efficiency and effectiveness toward experience is driving a focus on outcomes. We live in an outcome economy, and next-generation measures must be outcome-oriented and be able to influence performance in real-time. According to Joe Barkai, in his book *The Outcome Economy,* he explains that in an outcome-based economy, companies create value by delivering meaningful and quantifiable business outcomes, not just promises of outcomes. For example, Rolls-Royce no longer sells aircraft engines but rather assumes the responsibility for "time-on-wing." Airlines do not buy engines; they pay for engine availability or, you might even say, for lift power.

In the age of the Internet of Things (IoT), sensors are ubiquitous and provide enhanced capability to measure things that were not possible to measure easily. What if you could improve the performance of something you have been doing for over 20 years consistently at the same level with no changes, with next-generation measures? This is exactly how I improved my running distance and pace with a sensor (Lumo run) that could measure cadence (steps per minute), breaking (change in forward velocity), bounce (vertical oscillation), pelvic rotation, and pelvic drop. Notice these are all lead measures that influenced my performance, as I can hear the device provide real-time feedback in my ears while I am running. For over 20 years, I had just tracked lag measure of how many times a week I ran, with no impact on performance. From running sensors to smart yoga pants, to sensors in cars that promote safe driving, the next generation of measures are impacting performance in real time.

Project Management Artificial Intelligence (PI)

Project bots and digital assistants perform estimating, budgeting, and sprint management activities. As bots and sensors expand their understanding, new metrics that were not possible to capture will be used in the areas of quality, effort, performance, experience, learning, and change. New layers of meta-data will help in driving new metrics and insights that were not possible before. Remember, the example of decreasing the number of bugs above—AI-based system are linked, to know who made the changes to the code, and link the bugs reported to the line of code and the tasks related to it. Now, you have real-time, actionable measures to get to the root cause of the problem. Overtime PI will move from descriptive (what happened), and diagnostic (why it happened) measures, to providing predictive (what will happen), prescriptive (what action is required), and intuitive (learning and resolving issues before they happen) and impacting performance in real-time.

DEVELOPING MEASURE INTELLIGENCE

Use the following checklist of questions to reflect and develop measure intelligence. Pick and prioritize what to focus on over a period of time to enhance governance intelligence.

Objectives

■ How can we identify the right essential question that helps focus on the right thing in the sea of data?
■ What is the crucial question that needs to be asked?
■ How can we define success? What are we trying to achieve?
■ Do we know the business impact we are aiming for?
■ How can we use OKRs to clarify the purpose of measurement? Are our measures stakeholder- and customer-centric or self-centric?
■ What is the crucial question that customers care about?
■ How can we define and measure customer value?
■ How can we use metrics to increase customer satisfaction and experience?
■ Do our stakeholders care about the things we measure?
■ Are we using holistic and interconnected customer metrics such as net promoter score (NPS) and measures of customer satisfaction, customer success, and customer loyalty?
■ Are we using metrics effectively to improve our NPS and create customers?
■ How can we better measure customer and stakeholder experience and perception?

Key Results

■ Do current measures describe relevant current and past status and performance data?
■ Do current measures help predict and forecast future outcomes and results?
■ How can we utilize leading metrics that drive future performance and results?
■ How can we link and balance metrics from all elements of the DNA of strategy execution?

- Do current measures from other areas of the DNA cascade and align to strategy?
- How can we balance both, output and outcome measures?
- Do current measures provide insight and new ways of looking and dealing with existing challenges?
- Do current metrics reveal risks, barriers, and opportunities?
- How can we measure hidden value?
- How can we identify the key measure and related algorithm that can have a core impact on performance and results?
- How can we focus on behavior measures that impact performance?
- What counter-measures do we need to balance the behavioral impact of measures?
- How can we utilize metrics to assess overall impact?

Reporting

- How can we do a better job of communicating and illuminating value and results?
- How can we provide and emphasize context along with metrics to make sure metrics are interpreted correctly?
- How can we illustrate comparison and contrast to provide greater context?
- How can we evaluate and communicate causality and linkage of measures?
- How can we combine and communicate across to reduce provincialism and siloed approach to measurement?
- How can we improve the timing and timeliness of our reports and dashboards?
- How can we do better with separating the noise from the signal?
- Are our metrics credible to our customers and stakeholders—in how we capture, calculate, track, and report metrics?
- Are we using the right sensors, tools, and apps to collect, analyze, and communicate metrics?
- How can we improve the effectiveness of our reports, dashboards, and scorecards?
- How can we simplify and communicate more with less?

Action (Actionable)

- Are we using metrics just for autopsy or treatment?
- What can we do with the measures we track?
- Do current measures inspire appropriate action?
- Are our measures actionable?
- What actions can we take to counter the negative consequences of measures?
- Do current metrics help us make better decisions and prioritization?
- What would happen if we stopped measuring some things?
- What reports can we discontinue?

Learning (Feedback)

- Do current measures provide appropriate feedback loops?
- Do we get feedback on how to improve and adjust our measures?

- Are we tracking, measuring, and providing real-time feedback that drives the desired behaviors?
- How can we minimize the loopholes to game and manipulate metrics?
- How can we improve our data collection?
- Are we setup for rapid and frequent feedback?
- Are we learning and adjusting our measures?
- How can we improve our measures with balanced and holistic measures?
- What are we not measuring that we should be measuring?
- Are we building a metrics library over a period of time?

KEY TAKEAWAYS

- Identify and be clear about the purpose of measurement. Do you know the goal or objective you are trying to achieve? Do you know how your customers and stakeholders define success? Identify the crucial question that will help you focus on the right things.
- Collaborate with customers/stakeholders to define success and value and select appropriate metrics. It is important to engage them to understand their challenges and what value means to them.
- Develop and sharpen measurement intelligence in the five DNA strands of measure: Objectives, Key Results, Reporting, Action, and Learning.
- The Strategy Execution Measurement Framework provides an integrative approach to measurement. It helps to focus on the multiple dimensions and optimize the whole, rather than isolate some or overdose on others. It promotes a sense of ownership and linkage to outputs, outcomes, results, and impact.
- Make Ben and BoB your new besties. You can channel Ben (Benefit) and BoB (Benefit of the Benefit) by focusing on the distinction between outputs and outcomes. Ben is limited to the benefit of the output, whereas, BoB shows real value regarding business outcomes. Ideally, you need to strive for BobbyBen, who is bimodal and balanced and can leverage both aspects to achieve desired results and make an impact on strategic execution.
- Measurements drive behavior, and every measure has a consequence. The key is to identify the leading measures that drive the desired behavior.
- To start and apply the ideas of this chapter, define and measure success with the OKR framework and define balanced measures/metrics for Ben & BoB, and what leads to each of them.
- Seek feedback and adjust measures. Check and validate the metrics, collection mechanism, calculation formula, and outputs with stakeholders to see if this makes sense to them. The idea is to create a dialogue to keep inquiring what is valuable to them and the organization.
- Everything that may be of value cannot necessarily be measured. To harness the elusive nature of project management and PMO value, it is critical to manage perceptions and perceived value.

CHANGE

"If you focus on results, you will never change. If you focus on change, you will get results."

Jack Dixon

Leading Questions

- Why is change the catalytic element in the DNA of strategy execution?
- Why is it important to develop change readiness or risk being disrupted?
- Why is change hard and elusive?
- What are the core elements for holistic change?
- Why is it important to go beyond managing and leading change to change making?
- How to develop project management and PMO change intelligence?

It was only four weeks ago that Jonathan and his team had a go-live celebration party with great fanfare. The project sponsor and key stakeholders recognized the hard work and effort to get everything done and meet all the client requirements in time for the promised go-live date. In the first couple of weeks, as expected, his support team fielded a lot of calls from people who were trying to use the system. In the past week or so the calls have slowed down, but he has been hearing a lot of chatter regarding the system. The usage data that he was reviewing shows a considerable drop in system access. As he is getting ready for an urgent meeting with his executive sponsor, he is wondering what went wrong.

Julia was all excited to take on the new role of the PMO director six months ago. She had a good start with executive support to tackle a challenge that both executives and project managers faced to fix project reporting and provide good project data and status. After reviewing existing templates and processes for data collection, she rolled out a homegrown system to collect and report project data. She spent a lot of effort in rolling out the new process, with step-by-step instructions, demos, and multiple meetings. She can't understand why only about 40 percent of the PMs are actively using the system and updating data in a timely manner. She thought the slick dashboard interface would be a hit. Her initial excitement has been waning, as this is only one of the similar PMO implementation challenges she is facing.

What is the common thread in both these scenarios? What could they have done differently? In both the cases, they did not focus on change. They did not prepare their customers, end users, or the organization adequately, ready to adopt the changes that these systems brought. They seemed to do a good job on the execution or delivery side, but they failed to focus on the other half—change management—which is equally important for project success, particularly in today's world, where it is not enough to just deliver. Success depends on whether customers adopt and use the products, systems, or services. They have to be happy and satisfied with the experience and tell others and make it go viral, so the promised benefits and outcomes can be realized. And that may not happen without a holistic focus on change.

CHANGE IS THE CATALYTIC ELEMENT IN THE DNA OF STRATEGY EXECUTION

Strength in the change element can catalyze and accelerate strategy execution, whereas its weakness can derail and disrupt it. You can excel in all the other six elements of the DNA, which provide the structure, process, and capabilities for strategy execution, but without change, you are jeopardizing the adoption and intended outcomes of your initiative. The challenge is change is in the background and invisible, but it is the life force, and sustenance depends on it. Change must be surfaced and identified in the DNA and not treated as an add-on that is managed independently, as is often the case where change management is an afterthought.

Projects, programs, and PMOs succeed or fail based on how they understand, manage, lead, and leverage change from multiple perspectives—organizational as well as personal (psychological, emotional, neurological, and behavioral) change. Traditionally in the project world, change focus has centered on configuration management and is procedural. If you think you can win with process alone, you are mistaken. Our survey research over the past three years at the Projectize Group highlights the fact that PMOs spend 85 percent of their time on developing, documenting, standardizing, refining, training, and measuring process and methods. Less than 10 percent proactively plan for the impact of organizational and behavioral changes that these processes cause. This is one of the top reasons for lack of PMO effectiveness and buy-in. The adoption rate of PMO processes remains low.

Our own experience over the past 15 years of working with multiple fortune companies that have implemented project management and PMO practices shows that they have improved their project performance with better on-time and within-budget criteria. However, a common executive comment is, "Project performance has improved, but so what? End users and customers are not necessarily happy; they are not buying or using the product or the system!" The missing elements are the key questions of customer and end-user adoption, satisfaction with the product, or service of the project, which is all related to change. More and more, there is a realization of the need for a deeper understanding of change management, as well as a broader application and integration of change management. Also, there is a need to understand various aspects of organizational change management (OCM) that must be considered in concert with organizational project management practices.

Over the years, there has been greater awareness and recognition of the need to integrate project management and change management. "Regardless of the extent or maturity of organizational project management (OPM) in an organization, portfolio, program, and project management needs to increase the effective practice of change management inherent in the PMI foundational standards so that strategy can be executed reliably and effectively," according to PMI's practice guide, *Managing Change in Organizations*. The guide sets the practices, processes, and disciplines on managing change in the context of portfolio, program, and project management, and illustrates how change management is an essential ingredient in using project management as the vehicle for delivering organizational strategy.

Another aspect that needs to be emphasized is that traditionally the organizational, operational system default has been to limit, control, and manage change. In today's DANCE-world, it is the opposite: you risk disruption if you are not change ready. You have to not only accept change but encourage it. In the traditional world, you were conditioned to follow a plan; in the agile world, "responding to change over following a plan" is valued, according to the Agile Manifesto. One of the principles of agile is to "welcome changing requirements, even late in development. Agile processes harness change for the customer's competitive advantage."

Change readiness and acceptance is crucial and can be the catalytic agent, but is it easy?

CHANGE IS HARD

Imagine if you could press a few buttons, like one of those transformer toys that transform a robot into a tow truck, and accomplish change with a few flips. It would be simple if you were just dealing with procedural and mechanical aspects of change. Behavioral change is complex, with psychological, emotional, neurological, and other nuances.

You might think that people would change if their life depended on it. Behavioral change is hard even if it is a matter of life and death. In an eye-opening study, only one in seven heart patients can change their behavior, even when doctors tell them that they will die if they don't, according to a study referenced by Robert Kegan and Lisa Lahey in their book, *Immunity to Change*. If people don't want to change even when their life depends on it, imagine how hard it is going to be for the PMO.

A study of 14 contestants on the TV show *The Biggest Loser* found that despite getting world-class guidance, 13 of the 14 gained all the weight back. Imagine with world-class coaching and support they still could not keep the weight off. While they were on TV, they were super motivated and ate little and exercised a lot, but they could not sustain it after the cameras stopped rolling.

If individual change is hard, change at an organizational level is harder. We have to realize organizations don't change; it is the people who have to change. Studies show that two out of three transformation initiatives fail. According to a Towers Watson Change and Communications survey, more than 70 percent of change efforts fail and result in confusion, fatigue, and discouragement.

You may interview stakeholders, conduct pilots and receive feedback for a project that updates an in-house operational interface, for example, but it is still not easy for users to change their behavior and use the new system. How can you ensure project and PMO success when it is contingent on other people's behavior?

I am sure you will relate to the following list of project management and PMO challenges we repeatedly encounter in our practice:

- How to make them do what they are not excited about doing
- How to increase adoption of our project deliverables, or PMO processes & templates
- How to successfully implement a new governance process like stage-gates
- How to generate buy-in from stakeholders
- How to change the perception of the project or the PMO
- How to use the right measures that drive appropriate behavior
- How to influence executive behavior
- How to influence better prioritization and decision making
- How to increase the application and effectiveness of training programs

To deal with these challenges over the years, we have been working with clients to shift from focusing purely on configuration and procedural change, to a balanced behavioral and a multidimensional approach to change. Balancing a change-control mindset versus a change-readiness mindset; top-down versus bottom-up change, controlling versus self-regulating change, and operational versus a strategic perspective.

To implement these ideas in a practical way we have developed a PM Change Intelligence (PMCQ) framework. There are hundreds of books on change management and many different change models. There are several change adoption cycles in the change management literature starting with Kubler Ross model in 1969, to various modern-day variations. Among the popular change management processes are John Kotter's classic eight-step change process in *Leading Change:*

1. Establishing a sense of urgency
2. Creating the guiding coalition
3. Developing a change vision
4. Communicating the vision for buy-in

5. Empowering broad-based action
6. Generating short-term wins
7. Never letting up
8. Incorporating changes into the culture

Another common model is Prosci's ADKAR, which looks at five aspects of change:

- **Awareness** of the business reason for change—do people realize there is a need for change?
- **Desire** to engage and participate in the change—do they want to change?
- **Knowledge**—do they know what and how they need to change?
- **Ability**—are they able to implement the change at the required performance level?
- **Reinforcement** to make sure the change sticks—are they structurally encouraged to conform to the new way of doing things?

THE DNA STRANDS OF CHANGE

In our practice, we have used a combination of change models depending on the organizational context and culture. Over the years, we have culled our experience and fine-tuned our approach from a project management and PMO-specific standpoint and identified the strands or elements of change. These strands are organized as the 4A's + 4C's—the 4 A's of awareness, anticipation, absorption, and adoption (and adaptation) and 4 C's of customer, choice, communication, and connection which are essential to develop PM change intelligence (PMCQ).

Many of the classic models focus on different aspects of change adoption. The PMCQ is a broad and holistic approach that goes beyond adoption to awareness, anticipation, and absorption capacity. You can use the PMCQ to assess and develop your organization's change intelligence in a simple framework that combines a broad array of concepts, tools, and techniques based on research from the burgeoning field of behavioral economics, neuroscience, and contemporary change management literature. Figure 9.1 lists the DNA strands of change.

Awareness

"Awareness is the greatest agent for change."

Eckhart Tolle

In the busyness of execution and getting things done, it is easy to get caught up in the current tasks and issues of the day in front of you. The challenge is to develop change awareness—the awareness of the importance of change, awareness of its implications, awareness of biases and blindspots. Cognitive biases like confirmation bias, halo effect, false consensus bias, hindsight bias, loss aversion, framing effect, and others are tendencies to think in certain ways that can lead to systematic deviations

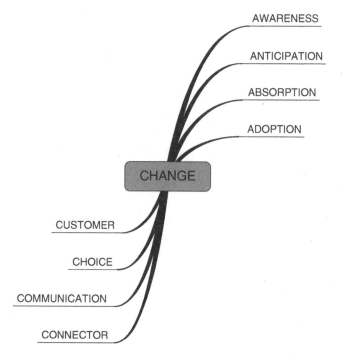

Figure 9.1: The DNA Strands of Change
Source: © J. Duggal. Projectize Group.

from a standard of rationality or good judgment, and are often studied in psychology and behavioral economics.

You have to develop awareness of various models and approaches to change management. There are various change adoption cycles in the change management literature, starting with variations of Kubler-Ross's Five Stages of Grief in 1969. After any change, there is bound to be a dip in productivity, when people are in denial, anger, bargaining, and depression, before acceptance of the change. The change cycle awareness helps to understand that resistance is normal, and before things get better they will get worse. It helps to be aware of prevailing perceptions and to have a good read on what is really going on. If people are nodding and in agreement with everything, they are probably not expressing the underlying resistance. It helps to cultivate awareness that change is hard, but you can be better prepared if you have a plan to minimize the dip as much as possible.

It is crucial to cultivate a culture of awareness—to have eyes and ears, to see and sense change at different levels, which promotes overall ownership, commitment, and accountability for change. Also, it is important to develop awareness of your culture, which can be an enabler or blocker, and drive behaviors that catalyze or impede change. The following questions can be used as a checklist of sensors to gauge overall change awareness:

- Is there awareness of the importance of change at various levels?
- Is there awareness of change management as a risk category, besides time, cost, scope, and other risk categories?

- Is there awareness of change adoption cycles at the individual/organizational level?
- Is there preparedness to deal with the various stages of the change adoption cycle?
- Is there awareness regarding prevailing cognitive biases and blindspots?
- Is there awareness regarding the social chatter and buzz regarding the changes this project or initiative is going to bring about?
- Have we done sufficient stakeholder assessment?
- Is there recognition of who needs to be committed to the change in order to be successful?
- Is there a good read on the barriers, neutrals, and allies?
- Is there awareness of the political climate for this change?
- Is there awareness of the cultural issues – the enablers and blockers, and the behaviors they lead to that can impact the change?
- Is there awareness of what do the stakeholders like about the current state, before the change?
- Is there awareness of what they dislike about the current state, before the change?
- Do they understand how the change will benefit customers and stakeholders?
- Is the organization structure appropriate for the change?
- Is there awareness of the prevailing perception of the project, program, or PMO?

Anticipation

"If the rate of change on the outside exceeds the rate of change on the inside, the end is near."

Jack Welch

Awareness helps to better sense and prepare for the consequences of the change. Anticipation takes it further to develop capabilities to anticipate what we cannot see currently, particularly the unintended consequences of the DANCE—dynamic, emergent, and unpredictable changes. The following questions help to develop a broader perspective to anticipate change better:

- Are there capabilities to anticipate the disruption caused by the DANCE (Dynamic | Ambiguous | Nonlinear | Complex | Emergent) elements?
- Are there capabilities and foresight to gauge the rate of change both internally and externally to avoid being blindsided?
- Are there adequate capabilities and competencies to sense, respond, adapt, and adjust strategy execution in a timely way?
- Is there awareness of limitations of traditional risk management and its inability to deal with the unknown, unknowables, or black swans?
- Is there awareness of PM processes and mechanisms that are limiting our anticipation capabilities and giving a false sense of security and faux-preparedness?
- Is there awareness of PM processes and mechanisms that could be the blinders that prevent us from seeing and anticipating the game-changing forces?
- Is there anticipation of possible organizational and business shifts?
- Is there anticipation of possible behavioral outcomes like anger, denial, resistance, rejection, and other related challenges?

- Is there anticipation of broader PESTLE factors—political, economic, social, technological, legal, and environmental impact?
- Is there awareness and humility that we cannot anticipate everything?

Absorption

One of the light bulbs that went on for one of the PMO directors we worked with was that for a while he didn't realize that when a project failed it was not due to execution issues. He and his team realized later that some of the areas that were impacted by the project did not have product readiness. The PMO was not aware and had not done enough to assess and prepare for change readiness and the absorption of the change.

When you think about change, you have to think about both the organizational and individual level. At the organizational level, you need to think about the volume (number of changes), velocity (the rate of change in a given time frame), the complexity, and impact of the change.

At the individual level, a change management expert, Donna Brighton, shares that terrific, talented people reach their capacity to absorb change, and they check out. Every person has their own "change sponge" that has a maximum amount of absorption. Both personal and professional changes decrease the change capacity. Employees become disengaged when they run out of capacity. All the leadership commitment, compelling cases for change, and brilliant change strategies in the world are irrelevant if you do not assess and manage change capacity. It is like continuing to pour water in a glass even after it overflows without assessing the absorption capacity. Instead of driving a tremendous amount of change in one phase, you have to think about team members can "absorb" all the major changes that depend on them. Water will still hit the floor from time to time, but it's due to the leader, and not because the change capacity, glass, or sponge of others isn't big enough.

Increase in absorption capacity can help build greater adaptability to change over time as individuals and the organization develop resilience to adapt to the changes. The following questions help to assess the change absorption capacity:

- Do we know our customers/end users' change absorption capacity?
- Do we have the capacity to handle the volume of changes that these projects are going to bring about?
- Do we have the capacity to absorb the velocity of projects being rolled out?
- Do we know the complexity and impact of the changes that these projects are going to bring about to the customers?
- Is there a process to assess the impact of changes and our customers' or end users' change absorption capacity?
- Is there an effective portfolio pipeline and prioritization process to assess capacity?
- Is there effective reprioritization of the portfolio?
- Is there an effective resource capacity planning process?
- Do we adequately retire or stop initiatives, programs, or projects?
- Do we have good data analytics capabilities to perform historical analysis for objective capacity assessment?

■ Are there allowances for ensuring that this (initiative or project) will not cause undue stress to the organization?

■ Are leaders willing to commit the right number of resources for the implementation of change?

■ Does the organization have the right skills/competencies to get the job done?

■ What is the degree of built-in awareness and resilience to deal with the unexpected?

You can also use change management tools like the DICE Framework from the Boston Consulting Group (BCG) to assess the absorption capacity and gauge the probability of project success before implementation. DICE assesses four elements and scores each project based on **duration**, **integrity** of team performance, **commitment** among senior leaders and the employees most impacted by the change, and **effort** additionally required from those employees.

Adoption

One of the core ideas of this book is that execution or implementation alone is not enough. It is only one side of the coin; adoption is the other side. Without adoption, implementation has no value. Next-generation project management creates awareness and an expectation that planning includes, not only tasks for installing the new systems, tools, and processes and the changes caused by them but also the activities for ensuring the adoption of these systems, services, or products. Next-generation project managers visualize implementation and adoption distinctly and plan for it. They recognize that unlike execution, adoption is an attitude, not an event, and they cultivate an attitude of adoption at all levels.

It is also important to highlight the distinction between adoption and adaptation. With rapid change, adoption may increase, for example, you may buy new tools or gadgets, or download more apps, but that does not mean you will use it, or adapt, to use it well and benefit from it. We are adopting technology faster than we can deal with. There is a lag between adoption and adaptation. G. Pascal Zachary from Arizona State University's School for the Future of Innovation in Society writes:

> Lag is the failure to adapt to changes.... The pernicious effects of lag persist longer than expected. The peril may be getting worse, in part because adoption of new digital tools is accelerating. Amazon, Twitter, Instagram—all seemed to have established themselves with the rapidity of a hurricane. In the distant past, adoption rates were slower, easing the challenge of overcoming social and cognitive lag.

Project managers and PMOs must focus on and prepare for both adoption and adaptation at the individual and customer level. (In this chapter, adoption is used, with the assumption that it includes adaptation as well). With adoption and adaptation, you have to realize that the organization itself cannot adopt and adapt; it is the people within the organization that have to change their behavior. Whether it is the adoption of the PMO's processes, tools, templates, apps, or services or it is the project's deliverables, individual adoption is the crucial last-mile problem that separates success from failure.

To understand individual change, we can reference a behavioral psychology mental model of the elephant, rider, and the path that was originally presented by psychologist Jonathan Haidt. According to the model, the rider is rational and can plan ahead, while the elephant is irrational and driven by emotion and instinct. This metaphor was made famous (and enhanced) by Chip and Dan Heath in their book, *Switch: How to Change When Change Is Hard*. They explain:

> Perched atop the Elephant, the Rider holds the reins and seems to be the leader. But the Rider's control is precarious because the Rider is so small relative to the Elephant. Anytime the six-ton Elephant and the Rider disagree about which direction to go, the Rider is going to lose. He's completely overmatched. You need to create a path that makes it easier to be successful.

They suggest directing the rider (head) to find ways to motivate the elephant (heart) and shape the path. For sustainable change, you have to engage and balance the head, the heart, and the hands (path). While the rider and the elephant (head and heart) are often focused on during times of change, it's the path that gets overlooked. You can tell people about why the change is important and why they need to change and even inspire and motivate them to change, but we forget to make it easy for them to change. This can be the opportunity for the PMO to shape the path and make it easy for the change to occur. Have you looked at the roadblocks in the path that could prevent easy adoption of the change? According to the authors, "What looks like a people problem is often a situation problem." For example, in the opening scenario of this chapter, Julia needs to review whether the new systems are accessible and available on all platforms easily. Perhaps they are not adopting and using the system because it is not easy to access, understand, or use. She can shape the path by removing the obstacles and making it easier.

The burgeoning field of behavioral economics (BE), pioneered by Daniel Kahneman and Amos Tversky's work, can help in understanding behavioral adoption challenges. According to the *BE Guide*, human decisions are strongly influenced by context, including the way in which choices are presented to us. Behavior varies across time and space, and it is subject to cognitive biases, emotions, and social influences. Decisions are the result of less deliberative, linear, and controlled processes than we would like to believe.

To apply the concepts of BE we have categorized four dimensions—the 4C's, or four focus areas, of change adoption, starting with the **customer**, the **choices** that can be architected, and ways to **communicate** to promote adoption, and utilizing **connectors** to influence and spread adoption.

Customer

If the customer is the center and everything else revolves around the customer, and if we want the customer to adopt our product, process, or service, shouldn't we start by understanding the customer? What does the customer need? What do they like and dislike? What motivates them?

Let's say I wanted to go fishing and I had never fished before. I am excited. I go and buy a fishing rod, and I am thinking that one of my favorite things in the world is brownies—come on, who in the world does not like brownies? So I go and buy the best brownies in town, and I hook a chunk of brownie and lob the line in the water waiting for fish to come. You might chuckle at this, but this is what we do without realizing we are feeding brownies to fish because we love them, without realizing that fish like worms.

The challenge is how to uncover what customers need. Just interviewing and talking to users at different touch points is not enough. You have to map the whole customer journey to understand customer experience, which is an approach used in design-thinking. For example, a hospital chain hired Ideo, a global design company, to improve the patient experience. In this scenario, the consultants didn't interview or survey the patients. Instead, the consultants registered as patients and slept inpatients' beds for 24 hours before proposing changes to existing procedures. You should do the same in the PMO. Live in the project manager's or customer's shoes. Feel their pain. See things from their perspective. Understand what is important to them. Then, you might be able to design products and services that meet their needs, and they may be more prone to accept and adopt them.

To hear the customer's voice, you can use another tool, Customer Impact Assessment (CIA) process, recommended by Christopher Frank in his book, *Drinking from the Fire Hose: Making Smarter Decisions without Drowning in Information*. The CIA consists of asking four questions:

1. How could this change negatively impact the customer?
2. How will the customer perceive this change?
3. How are you going to manage this change for the customer?
4. How are we going to track the impact of this change, and how precisely are we going to measure its progress?

A tool that we use in our practice is an empathy map (discussed in Chapter 3) to map what the customer or stakeholder sees, hears, thinks and feels, says and does, and what is their pain and gain. Figure 9.2 illustrates the empathy map of project managers regarding Julia's new reporting template that she is rolling out.

You can see how Julia could have used the empathy map to assess and address the underlying issues and be better prepared before rolling out the template. You can use it to surface and decode the wants, needs, likes, dislikes, motivation, and expectations. Your goal is to see how you can increase their likes and decrease their dislikes—how you can make it easy for them and tap into their motivations that propel them to adopt.

The challenge for the PMO is how you can help them love what they hate, and how to help them to do what they can't. In their book, *Influencer: The New Science of Leading Change,* Joseph Grenny and Kerry Patterson provide a framework for influencing change from a personal, structural, and social standpoint providing ideas, tips, and

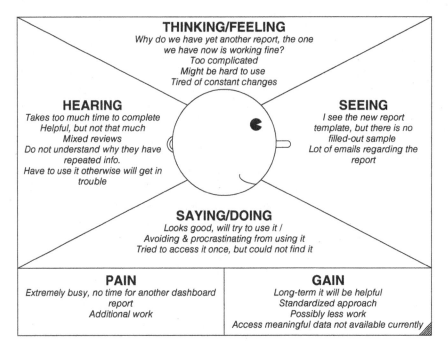

THINKING/FEELING
Why do we have yet another report, the one
we have now is working fine?
Too complicated
Might be hard to use
Tired of constant changes

HEARING
Takes too much time to complete
Helpful, but not that much
Mixed reviews
Do not understand why they have
repeated info.
Have to use it otherwise will get in
trouble

SEEING
I see the new report
template, but there is no
filled-out sample
Lot of emails regarding the
report

SAYING/DOING
Looks good, will try to use it /
Avoiding & procrastinating from using it
Tried to access it once, but could not find it

PAIN
Extremely busy, no time for another dashboard
report
Additional work

GAIN
Long-term it will be helpful
Standardized approach
Possibly less work
Access meaningful data not available currently

Figure 9.2: Empathy Map of Project Managers who have to use the new Reporting Template

examples to increase motivation and ability from each aspect. To help them love what they hate, they describe four tactics to make pain pleasurable:

1. **Allow for choice.** You can never hope to engage people's commitment if they don't have permission to say no. Instead of using top-down compulsion, the PMO should provide an up-front choice that they can continue to do it the old way, if the new change is not beneficial to them.
2. **Create direct experiences.** Let people feel, see, and touch things for themselves. There is no substitute for direct experience. Engage them in interesting demos; let them try it out and experience the system or product for themselves.
3. **Tell meaningful stories.** You can create vicarious experiences by sharing meaningful stories that resonate with stakeholders. Stories can evoke ephemeral emotions that motivate change. You can build a PM repertoire of stories that can be used to inspire new insights for the project or PMO.
4. **Make it a game.** Gamification is engaging, fun, and exciting and works. You can use contests and competition to involve people and make pain pleasurable. Keeping score helps with continuous improvement and learning. The challenge is to identify and tweak the measures that drive the right behaviors.

To help them do what they can't, you have to not only help them learn the new skills, but you have to help them practice and apply the new behaviors. Make sure that practice involves realistic conditions, coaching, and feedback. Break the vital behaviors into smaller actions that allow people to judge how well they're doing. And finally, help others practice how they'll recover from setbacks should they fail in their early attempts. Be sure you help people develop not just the technical or interpersonal skills they need to succeed but also the intrapersonal skills. Engage them in practice in addressing emotions that might undermine their attempts to change.

Measurements drive behavior. The key is to know what measures drive what behavior and focus on them. When you choose relevant measurements and show their effects, you can experience a sustaining impact on behavioral change. For example, one of the PMOs found that project teams were spending too much time in too many ineffective meetings. The employees also were in the habit of taking work home at night because they didn't have enough time during the day to complete it. The PMO took on a project to change that behavior. First, they measured the time and cost of each employee who attended regular meetings. Then the PMO visually displayed this data and showed how it impacted productivity. When the project results were presented, the project managers felt confident that people would change their behavior to conduct fewer, shorter, and more effective meetings based on data and measurements.

An in-depth discussion of measurements and ways to address measurement-related questions are covered in Chapter 8.

The following questions can be used as a checklist to apply the ideas discussed above and further develop PMCQ from a customer standpoint:

- How can we effectively connect and empathize with our customers and end users to better understand their needs and expectations?
- How can we better understand the likes or dislikes of our customers?
- How can we better understand the motivations and aspirations of our customers?
- What tools can we use to hear the customer's voice?
- How can we better understand how the customer will perceive this change?
- Do we allow for the choice to say no?
- How can we utilize opportunities to provide direct experience for the proposed change?
- How can we effectively use incentives to cause change?
- How can we use gamification to engage customers/stakeholders and promote input, improvement, and right measures?
- How can we better identify the vital customer behaviors and break them into smaller actions?
- Are we measuring the right things to drive the right behaviors?
- How can we track, measure, and provide real-time feedback that drives the desired behaviors?
- How can we track the impact of the change on our customers?
- Is there a process to help customers/stakeholders recover from setbacks?
- How can we better engage our customers/stakeholders to generate buy-in?
- How can we instill appropriate triggers and desirable habits?
- Do we understand what drives customer/stakeholder success?
- How can we better monitor and manage customer/stakeholder perception?

Choices

The choices that you provide to the customer help in paving the path and help to design the structure toward desired outcomes. Choice architecture is based on the work of Thaler and Sunstein, described in their 2008 book, *Nudge*. They provide a framework of different ways to nudge people based on different options like order—the relative position in which choices are presented (high-medium-low budget options), capitalizing on default options—choices that don't require action (e.g., default to opt-in for organ donations), or accessibility—easy to see, reach, choose, or think (e.g., easy accessibility of systems or visibility of actionable information). Much of this theory and terminology was first established as heuristics by Daniel Kahneman and Amos Tversky. Thaler and Sunstein essentially equate these "heuristics" as "nudges."

Table 9.1 lists some examples of nudge heuristics relevant to PM and PMO choice architecture and communication (adopted from the Nudge Theory Toolkit; for more, – go to http://www.businessballs.com/nudge-theory.htm#nudge-toolkit).

Checklists

Structured checklists can also help pave the path toward greater adoption. Part of project and PMO governance is to provide standardized checklists. These checklists can also be a behavior change vehicle if done right. Atul Gawande, a surgeon, and best-selling author, in his book, *The Checklist Manifesto,* explains how checklists can help with memory recall and clearly set out minimum steps necessary in a process. Good checklists are explicit. They offer the possibility of verification but also instill the discipline of higher performance.

Gawande discusses a study using a basic checklist to prevent central line infections. Checklists, he notes, established a higher standard of baseline performance. They didn't force anyone to use the checklist but started by gathering data of infections, and having insurers pay small bonuses for participation. They also required senior hospital executive participation and involvement and gave responsibility for the checklist to least powerful people in the process (the nurses).

According to Gawande, ineffective checklists are vague and imprecise, too long, hard to use, impractical, made by desk jockeys without functional knowledge of the field (which is often the complaint about the PMO staff), try to spell out every detail and turn brains off. Effective checklists are efficient, to the point, can be used in the most difficult situations, and, above all, are practical.

The following questions can be used as a checklist to apply the ideas discussed above and further develop PMCQ from a choice standpoint:

- How can we use relevant nudge tactics (anchoring, framing, priming, social norms, comparison, and contrast) to design and provide appropriate choices to achieve desired behaviors?
- How can we implement and improve checklists as behavior change vehicles?

Table 9.1: Nudge Heuristics Relevant to PMs and PMOs

Nudge	Usage Examples
Anchoring and adjustment	Give the audience factual comparisons and references that are relevant to them. Publicize statistics and facts about things that distort people's thinking and are misleading influences. When providing project reports and status updates, use previous projects and data as anchors to compare and contrast. Maybe your project status is not as bad in comparison to the other project.
Loss aversion	Focus on gains and improvements. People don't like losing benefits or incentives they already have. Avoid presenting situations as "taking" things from people. Clarify actual values of possessed things. For example, remind them that if they do not submit timely PMO information, they will lose the privilege of no-meeting Fridays.
Status quo bias and inertia (default bias)	Use status quo and defaults to helpful positive effect. Make it easy for people to make helpful choices. Make it difficult for people to use faulty thinking and make unhelpful decisions. Prefill default options on project templates to make it easy and drive right choices.
Framing	Design communications so that choices are positioned and explained positively—relevant and clear to the audience. Use words carefully. Understand and focus on what your communications mean to people, rather than what they mean to you. Orient communications according to received meaning.
Priming	Help people visualize positive actions and outcomes. Encourage people to consider, explore, and assess steps and strategies—how they can do things. Help people think and decide. Avoid thinking and deciding for people. Educate and inform people, especially as to relevant causes and effects.
Positioning	Understand how the positioning of things affects people's engagement with them. For example, placement of certain project data on a dashboard in a certain to order to call attention, information radiators with current project updates in strategic locations, or signs in frequent meeting places.
Limiting	Helpful options can be more appealing if they seem limited by time or availability. For example, have only limited time availability for exclusive access to pilot the new project portfolio management (PPM) tool.
Accessibility	Find ways to increase the percentage of your audience who will see or experience the intervention and the number of times they see or experience it. Make sure the new systems or templates you want them to use are accessible and available on all platforms easily.

Source: Adopted from http://www.businessballs.com/nudge-theory.htm#nudge-toolkit

Communication

In today's world of media overload, everybody is drowning in ever-increasing stimuli and information. To connect you need to incorporate vivid experiences that activate multiple brain receptors and frame memorable and sticky messaging. For example, a PMO in a professional services company sought to update its methodology. This involved creating new workflows and processes. The proposed change was important, and the PMO knew they had to communicate it to the employees in a relevant way. They needed to grab the users' attention and reach them on an emotional level to stimulate behavioral change. Users were invited to a bonfire to burn all the old process manuals. That successfully caught their attention in a visual and emotional way and motivated their change.

The following questions can be used as a checklist to apply the ideas discussed above and develop PMCQ from a communication standpoint:

- How can we provide a clear and unambiguous vision/future state related to the change?
- Are we providing adequate education about the change and the context and need for the change?
- How can we communicate the change in an engaging and exciting way that grabs their attention and creates an emotional connection?
- Is our message inspiring and has a transformational purpose and meaning that resonates and motivates?
- Is there consistent marketing and communication on the need for change?
- Is the timing right for greater connection and resonance of the message?
- How can we utilize storytelling techniques to drive change?
- How can we incorporate relevant nudge techniques of anchoring, framing, priming, social norms, comparison, and contrast in our communications?
- How can we create a safe outlet for feedback including reactions, concerns, and comments?
- Do leaders demonstrate a commitment to the change through actions as well as words?
- How can we better communicate the expectations related to the change and what the definition of success looks like?
- How can we ensure that the success measures for the change are well defined and understood?

Connectors

You have probably experienced that if an executive or senior leader needs to stop an initiative, program or project, it is relatively easy to do it from the top with an edict. However, when they have to start something new or cause any kind of change, they cannot do it on their own. They need the connectors—many agents at different levels that are infecting other and spreading the positive virus. These are typically the opinion leaders who hold clout and respect that people pay attention to. The top-down change can be catalyzed with the effective use of the social power of the connections in the network. For more on this, please see Chapter 7.

Identifying and Amplifying Positive Deviance

Another related aspect of connectors is to identify the positive deviants, which can be pivotal for change. Traditionally, deviance is referred to as intentional negative behavior. In the past few years, behavior scientists have been studying deviant behavior that is positive in various societies and cultures and also extending some of the findings to organizations. Behavioral scientists Gretchen Spreitzer and Scott Sonenshein define organizational positive deviance as "intentional behavior that departs from the norms of a referent group in honorable ways."

Positive deviance (PD) is an approach to personal, organizational, and cultural change based on the idea that every team, community, or group of people performing a similar function has certain individuals (the "positive deviants") whose special attitudes, practices/strategies/behaviors enable them to function more effectively than others with the same resources and conditions. Because positive deviants derive their extraordinary capabilities from the identical environmental conditions as those around them but are not constrained by conventional wisdom, positive deviants' standards for attitudes, thinking, and behavior are readily accepted as the foundation for profound organizational and cultural change, as they are also often the connectors.

Positive deviance is based on three criteria—voluntary behavior, a significant departure from the norms of the referent group, and honorable intentions. While implementing new processes and methods, PMOs look at introducing best practices from the outside. These practices have a hard time in gaining acceptance and adoption and are often rejected because these are foreign practices and perceived as yet another irrelevant trend of the day. Instead, PMOs need to understand the power of PD and identify positive deviants from within the organization who are delivering successful projects by using and applying methods or tools that are not the norm in the organization. The PMO's role becomes to identify the positive deviants and amplify and support their practices and spread the positive practices from within. The resistant PMs and team are more open and prone to identify with practices that are promoted from within by positive deviants, especially if these deviants are the connectors who are successful and respected within the organization than coming from the outside.

The following questions can be used as a checklist to apply the ideas discussed above and further develop PMCQ from a connectors standpoint:

- How can we effectively engage different levels of stakeholders?
- How can we better identify and analyze the underlying stakeholder network and relationships?
- How can we identify the influencers and opinion leaders who people look to for making their decisions?
- How can we better identify the positive deviants?
- How can we encourage and incentivize the influencers and positive deviants to share their commitments and actions?
- How can we provide a platform and amplify the message of the influencers and positive deviants?

HOW TO DEVELOP CHANGE INTELLIGENCE

Use the above questions in each of the 4A's + 4C's areas to determine your change intelligence. You can reword the questions in a couple of different ways—to do a self-assessment to determine whether that particular aspect exists or not. The greater number of positive responses the higher degree of change intelligence. You can also use the questions to discuss your strengths and gaps and ways to improve these areas to develop PM/PMO change intelligence. These questions can be assessed from a multidimensional perspective—the PMO itself, project/program team, sponsors, executives, project/program managers, team members, and other relevant stakeholders.

Dealing with Change Resistance

Whether your project is about implementing a system, handing over a completed building site or rolling out a project management office process, you are bound to encounter resistance to change and acceptance. "Even after 10 years some people have not fully accepted the transition, they are still resisting it, they are forced to use it, but they don't accept it," admits Delores Johnson, a PMO manager, from a public-sector organization, about one of the major projects that she has been involved with. Let's say you are implementing a system or completing a new facility. The end users are going to resist it because they have a lot invested in the current way of doing things. They may feel that it is not fair and justified how the project is going to affect them. The people in the new facility might complain because they didn't know what to expect. They were not involved from the beginning and believe some aspects of the project are not justified and don't agree with certain facets of the project.

Remember the idea of fair process and procedural justice introduced in Chapter 2. The idea based on the work of two social scientists, John W. Thibaut and Laurence Walker, was further developed as a management concept by W. Chan Kim and Renee Mauborgne. They discovered that people would commit to a manager's decision and change—even though they might disagree with it—if they believe that the process the manager used to make the decision was fair and justified.

Fair process is based on three mutually reinforcing principles: the 3 E's—engagement, explanation, and expectation clarity. We have added another element in our practice the fourth E—empathy.

Engagement means engaging stakeholders throughout the project lifecycle and proactively seeking their input, particularly in aspects of the project that will affect them most. Engagement provides a sense of confidence that their opinions have been considered.

Explanation details the decisions and makes sure the stakeholders understand the key points. You cannot assume that decisions are self-explanatory or straightforward. More importantly, explanation should also provide the background of why project decisions were made. This provides people with the context as they try to assimilate and adapt the changes from the project.

Expectation clarity describes the "new rules of the game." It requires clarification of the expectations and consequences brought about by the project.

Empathy involves putting yourself in the shoes of your stakeholders to understand and feel the pain that the change is going to bring about. This helps you better plan to make the change process fair and just and connect with them from their perspective.

Ms. Johnson remembers that "there was no explanation of why and what was the benefit or impact. Only two representatives were sent for training from each department, but they were not able to involve or disseminate important information to the rest of the group." The four E's of project justice—engagement, explanation, expectation clarity, and empathy—applied throughout the project life cycle can ensure that these questions are addressed adequately and help to reduce resistance. Remember, buy-in and acceptance are directly proportional to engagement. Conversely, estrangement can induce powerful resistance to change.

How to Get Results without Being a Nag

Once again, your project report is due today, and you are still waiting for status updates from your team members. As you get ready to write yet another nagging e-mail, in bold letters with red underlined due dates and timelines, you wonder: Why do you have to go through the same cycle time and time again?

Getting your team to do routine administrative tasks is akin to the challenge of having your children pick up after themselves or doing their homework on time. These tasks are typically considered tedious and trivial compared to other more pressing project priorities. Nagging e-mail reminders do not necessarily help because after a while people get desensitized. Task alerts and project collaboration tools can be helpful, but what do you do if people don't use them? There are just too many alerts and e-mails that get lost in the shuffle. The offenders think this is the norm and others are doing it too, so it is okay to be tardy. So—how do you get people to do routine tasks without being nagged?

Start by observing the individuals who do turn in their reports on time. For them it is a habit—they have programmed themselves to complete these tasks without much thought. The trick is making routine tasks habitual and easy so they become automatic practices. Dan and Chip Heath explain how habits become behavioral autopilots and offer valuable tips such as tweaking your environment and setting action triggers to build habits.

The following are ideas and strategies to cultivate habits, organized in the mnemonic **HABIT**:

Habitat—People are comfortable in their current habitat and ways of doing things. To get them to change and to adopt new habits, you have to tweak the environment. Alain Gervais, PMP, a project manager of 20 years from Ottawa, Canada, recently had a breakthrough in getting his team to send him weekly updates. Changing the environment by simplifying and automating the existing reporting mechanism promoted a culture in which timeliness became important.

Act—Often, tasks don't get done because people don't have everything they need to act. It may be that the process to complete the task is too complicated or not well understood. Ask questions about your team members: What do they need to act? Do they have the access? Do they need training? Do they need support?

Benefit—Why should they make an effort to change? Explain and emphasize the benefit of timeliness to the team or the overall project needs. They need to understand the context and the consequences. For example, if they complete their project updates accurately, they can skip the status meeting and have additional time to work on other things.

Incentive—Offer rewards or recognition for timely submittal or completion of administrative tasks. You can introduce an element of fun and excitement around routine tasks by creating competition and contests and celebrating success.

Triggers—Not just automated alerts and pop-up messages, action triggers that are specific, and visual work to program yourself or your team members to take action. For example, "Tuesday morning coffee update" can remind you to complete a project report, or an "okay to use electronic devices" announcement on the airplane could be a trigger to work on your expense report after a business trip.

Be clear about the expectations and consequences. People don't do what they are supposed to do because there are no consequences. Changing habits is not easy. To sustain habits, it is important to track, measure, and report. After all, measurements drive behavior.

From Managing Change to Changemaking

The prevalent perspective of change particularly in project management relates to managing, controlling, and monitoring change. There is a lack of ownership and responsibility for the success of the change. There is a fundamental shift that needs to happen from managing and leading change toward the idea of owning and making change happen. Traditionally, project managers manage change to deliver the project. However, the responsibility and accountability for the change taking place is detached. There is no direct sense of ownership of the change as a result of the project. Execution and delivery are measured separately from results and outcomes. A project manager's perspective is that it is not his or her responsibility whether the change happens or not, and whether the end users change, adopt, and use the products, systems, or services of the project. As a result, there is an inherent dichotomy between delivery and adoption or implementation and the use of the product or service. Project managers are not responsible for the end use or success of the initiative, product, or service. There is a need to shift this perspective and bring the pieces back together. The next generation of project management needs to emphasize the ownership and accountability of ensuring the change takes place to position and enable the achievement of the desired benefits and outcomes. The next-generation PMO needs to inject greater change intelligence into the organizational culture to drive a holistic and integrative approach, rather than a reductionist and procedural view.

Whether you are a project manager or program manager, you will benefit from a shift in perspective from managing and leading change to change making. *Changemaker* is a term used in the context of social entrepreneurship. A changemaker is one who desires change in the world and, by gathering knowledge and resources, makes that change happen. It is a term popularized by the social entrepreneurship organization Ashoka. I believe this term is much needed in project management, to cause a shift and bring about a different perspective. A changemaker isn't someone who simply manages change and wishes for change; he or she makes change happen. Changemakers have the ability to transform; they influence the outcomes through responsibility, ownership, and determination.

A post from *Observations of a Changemaker: Lessons and Stories from Future Changemakers* explains:

> There is one main difference between changemakers and the rest of the world. Most people desire change; many know what they would like to see different in the world, and some even know how it could be done best. There is a higher class of people who act to see their change happen; many of these fail. However, this is where changemakers differ: they *make* their change happen. Using a combination of knowledge, resources, and determination, they push through until their dream becomes truth, and then push some more.

It is important to distinguish between managing change versus making change. Table 9.2 summarizes the key points of distinction between managing change and making change.

Table 9.2: Managing Change versus Changemaking	
Managing Change	**Changemaking**
Manage and lead change	Own the change and make change happen
Focus on execution and delivery	Focus on end-user/customer adoption and experience
Focus on goal and end-date	Focus on goal with a long-term gaze
Extrinsic approach	Intrinsic approach
Oversight, control, and alignment	Insight, accommodation, and adjustment
Ownership and accountability of deliverables and outputs	Ownership, accountability, and determination to see change through and focus on results, outcomes, and impact of change
Limit and control change	Expect, embrace, and encourage change
Change weary	Change ready
Procedural focus	Behavioral focus
Manages and leads from the top-down	Manage and lead from the bottom-up
Hands (tactical) and head (strategic)	Balance of hands, head, and heart (behavioral and emotional)

CHANGING THE WORLD WITH CHANGE INTELLIGENCE

Imagine you were fundamentally trying to change age-old practices in an industry. You had to communicate your plans and sell your project to strong-minded executives in a very close-knit traditional industry. You are not only an outsider but as a woman in a male-dominated industry; you are not even allowed on the premises. Would you dare to even try it? Well, the story of Temple Grandin is inspirational and instructive because she changed the world of animal husbandry and slaughterhouses. Today, over half the cattle in North America are handled in humane systems she designed. How did she do it? As I watched the HBO documentary of her story, it became clear how she inherently cultivated change intelligence and utilized the 4A's + 4C's to cause change.

Temple developed an acute sense of awareness and anticipation from her childhood due to her autism. From early on, she has claimed that her autism created a special relationship between her and animals—based on a deep understanding at the sensory level. This comes in part from her keen ability to see the world in vivid images.

The customer in her case was the animal as well as the cattle and slaughterhouse owners. The close connection Temple feels with animals was central to her research and design. Temple would evaluate each handling facility by visualizing it from the animal's standpoint—including sometimes crawling through gates and chutes to identify what sounds, sights, or smells might cause them to be fearful and balk. Her other customer, the owners, did not care about humane treatment of animals who were going to be slaughtered anyway. So she had to present her ideas in a way that they could connect from a savings and efficiency standpoint to get them to change their mind.

She focused on choice by incorporating both the animals and the owners' perspective into her designs. For example, she designed curving, serpentine walkways that both prevent cows from seeing the slaughter up ahead and panicking. The new design gives the animals the sensation that they are coming back around the same way they came in. Temple highlighted and removed objects and design features that could easily spook animals. She started with minor retrofits to existing facilities and then evolved into larger designs of high-volume animal processing facilities and a broader approach that would create wide adoption.

Temple realized that just inventing solutions does not cause change. She learned to actively use communication and connectors. She focused on the efficiencies and communicated the payoffs and business benefits at play. She identified the allies and connectors who could help her to communicate the change. She found people in the slaughter facilities who trusted her and helped her articulate the problem by contributing data on behaviors of animals.

In the opening scenarios, Jonathan and Julia were low in their PMCQ as they did not have change awareness and the anticipation of the impact of the changes that their project and PMO would bring about. They did not assess the absorption capacity or focus on the 4 C's of adoption—to understand what their customers and stakeholders really wanted and needed, how they could focus on choice to make it easy and accessible, and communicate in a way that it resonates and motivates them to change, and use the social power of connectors to catalyze change at different levels.

Change is the catalytic element of the DNA of strategy execution, and when you think about change from a project and PMO standpoint, you can start with extrinsic factors to shape the path and make adoption easier. To achieve sustainable change, you can also focus on the elephant and rider. If your PMO can create the right conditions by igniting and sustaining the intrinsic desire for meaning and purpose and the impact of the work itself, you can catalyze strategy execution.

> "If you want to build a ship, don't drum up the men to gather wood, divide the work and give orders, Instead, teach them to yearn for the vast and endless sea."
>
> *Antoine De Saint-Exupery*

If you continue to develop change intelligence and some of the ideas described in this chapter, you will be perceived as change-ready—an enabler, not a blocker of change. Ready to adapt and adjust to the DANCE and the changing reality. Also, focusing on change as a catalytic element of the DNA cultivates a change action mindset, which is about driving action. One simple way to practice that is to stop thinking, "*What do I need to **manage** today?*" Instead, start with, "*What do I need to **change** today?*"

KEY TAKEAWAYS

- Change is the catalytic element in the DNA; either it can catalyze strategy execution, or its weakness can slow and stall it.
- Projects, programs, and PMOs succeed or fail based on how they understand, manage, lead, and leverage change from multiple perspectives.
- Traditionally, there is more focus on procedural change in project management; there is a need to shift toward behavioral change.
- If you are still operating with the default PM mindset to limit, control, and manage change, you risk being disrupted. You need to do the opposite—expect, embrace, and encourage change.
- To develop change intelligence understand and practice the 4A's + 4C's. The 4A's of awareness, anticipation, absorption, and adoption (and adaptation) and 4C's of customer, choice, communication, and connection which are essential to develop PM change intelligence (PMCQ).
- Map your organizational culture and identify the enablers and blockers, that drive certain behaviors that catalyze or impede change.
- There is a lag between adoption and adaptation. Project managers and PMOs must focus on and prepare for both adoption and adaptation. Make it easy and remove any obstacles for adoption and adaptation.
- Next-generation project managers and PMOs are changemakers; they make change happen. They have the ability to transform; they influence the outcomes through responsibility, ownership, and determination.

10
LEARN

"The illiterate of the 21st century will not be those who cannot read and write, but those who cannot learn, unlearn, and relearn."

Alvin Toffler

Leading Questions

- Why is learning the key to adaptive agility and survival?
- Why are traditional project management and PMOs not setup for learning?
- What is the next-generation perspective needed for learning?
- How do we rethink failure and make it survivable?
- What are the seven elements of designing a learning organization and PMO?
- How can the PMO become the curator of organizational practices to create meaning from noise?
- Why should learning be an essential measure of progress?
- How do we develop learning intelligence?

What if your companies' DNA was organized to take all the outputs it generates, and use them as inputs? How could you take the outputs, whether it is products or all means of data generated from the business and use it to learn, evolve, and innovate?

This is exactly what companies like Amazon, Google, and Facebook are doing well. The growth trajectory of these companies is pumped by a continuous feedback loop of learning, experimentation, and innovation. In the case of Amazon, they evolved from selling online books to a leading retailer of almost everything, to electronic devices, web services, media, food, and a host of other related products and businesses. These companies were born and flourished in a fast-changing DANCE and disruptive world, so they inherently know that learning is the key to organizational agility and survival.

Table 10.1: Organizing to Execute versus Organizing to Learn	
Organizing to Execute	**Organizing to Learn**
Confirm and follow rules	Experiment and solve problems
Objective—choose among defined options	Empirical—experiment through trial and error
Learning before doing	Learning from doing
Did YOU do it right?	Did WE learn?
Separate expertise	Integrate expertise
Drive out variance	Use variance to analyze and improve
Works when path forward is clear	Works when path forward is not clear

Source: Adapted from *Rotman Magazine*, Winter 2015, Amy Edmondson, Thought Leader Interview, by Karen Christensen.

If you look back at the history over the past century, the companies that have survived in the long run like IBM, GE, and Procter & Gamble have all survived because they also have learning and evolution in their DNA. Like IBM, which has evolved from products to services in the 1990s, to cloud, mobility, and security in recent years, unlike Blackberry, Motorola, Nokia, and many others that did not learn and sense, respond, adapt, and adjust quickly enough.

According to Richard Foster of the Yale School of Management, the average lifespan of a company decreased from 67 years in the 1920s to just 15 years today. He estimates that 75 percent of S&P 500 companies will be replaced by new ones by 2027. To survive, learning must be a part of the organizational DNA, so it can evolve and mutate based on changing conditions.

But the challenge is that most companies are not organized to learn; they are set up to execute. As Harvard Business School Professor Amy Edmondson points out in Table 10.1, the difference between the two are the characteristics necessary to cultivate learning into the DNA.

TRADITIONAL PM AND PMOs ARE NOT ORGANIZED FOR LEARNING

Project management and PMOs, as you can imagine, are typically setup to execute and have more of the characteristics of the left side Table 10.1. The traditional mindset is that you are supposed to follow the rules, do it right the first time, and drive our variance, and failure is not an option. This does not cultivate the right conditions for learning, which requires the opposite environment of experimentation and trial and error.

Of course, project management and PMOs always emphasize lessons learned, continuous improvement, and maturity, but they continue to struggle, as they are not setup for success.

Need for a New Perspective

Cathy has been preparing for her upcoming project closeout. One of the items on her checklist as the project manager is lessons learned. She has it scheduled on the calendar and invited key project stakeholders to the lessons learned meeting. She has asked her team to send her ideas for the presentation. She gets a few e-mail responses, and a couple of days before the meeting she works on her slide deck. She is happy that the PMO has provided a template for the lessons learned. As she works on her presentation, in the beginning, she struggles with what to list, but then she can remember all the typical challenges and lists things like more planning, more resources, need for sponsor involvement, timely risk management, greater communication, and team building. She delivers her lessons learned presentation. There are a couple of questions, and she provides good examples. Everybody seems to be happy. Now, she can check this item off, and officially close out the project and look forward to the celebration party!

Sound familiar? Well, you could cut and paste things like more planning, more resources, need for sponsor involvement, timely risk management, greater communication, and team building, and you don't need to do another lessons session because every project has very similar lessons.

So what are the problems the above scenario highlights?

The lessons learned was a cursory exercise, so that she could check it off that it was done as a part of the PMO process. There was no real reflection or lessons here. Why did she wait until the end of the project, how come lessons were not documented throughout the project? There were probably many lessons lost as they were not captured at the right time. Once they are filed, they are seldom looked at again; people don't even know where to find them. Even if they find them, the lessons are generic bullet points without any context, and not actionable. You might review all the project artifacts and documentation, but they may not tell the whole story, particularly the wealth of tacit knowledge and insights that are not easy to harness.

What about Continuous Improvement and Maturity?

Continuous improvement of process is good, but it is not enough. Improve not only the process but also the work. Not just improve but evolve and innovate. Continuous improvement is a single loop, you improve the same thing, over and over, which may not be relevant in a changing world. Continuous innovation is a double loop, where you learn and evolve to create something new and better.

As we have discussed in previous chapters, you can be mature but not necessarily smart. You may have well-defined, standardized, repeatable processes, but in a

DANCE and changing environment, they may not necessarily be applicable. You are organized to execute, but not to learn and evolve. In fact, it can promote a false sense of security that you have reached a certain level of maturity and have an advantage.

To address this, you need to rethink and ask the right questions to cultivate the operating system (OS) of the organization for learning: How do we utilize agile principles to reflect on how to become more effective and adjust based on empirical observation? How do we treat failure? Do we encourage experimentation and trial and error? How do we make failure survivable?

LEARNING FROM FAILURE

> "Good judgment comes from experience; experience comes from bad judgment."
> *Nasrudin, thirteenth-century wise fool of Sufi lore*

In one of the reflective moments in his book, *The Lean Startup*, Eric Ries at one point shares a painful moment of self-doubt. The entrepreneur has just recounted the very early days of one of his startup companies, which scrapped its initial strategy and product after realizing they were fatally flawed. Thousands of lines of code, which Ries himself had written during the past six months, is thrown out as the company pivots to a new strategy based on customer feedback. He writes:

> Would the company have been just as well off if I had spent the last six months on a beach sipping umbrella drinks? Had I really been needed? Would it have been better if I had not done any work at all?

> There is … always one last refuge for people aching to justify their own failure. I consoled myself that if we hadn't built this first product—mistakes and all—we never would have learned these important insights about customers. We never would have learned that our strategy is flawed. …

> For a time, this "learning" consolation made me feel better, but my relief was short-lived. Here's the question that bothered me most of all: if the goal of those months was to learn these important insights about customers, why did it take so long?

Ries ultimately realizes that he had wasted time doing things that didn't uncover what created value for customers. They had followed agile development methodologies that prized efficiency but had used them in the service of building out product features it turned out no one wanted—which in turn were an outgrowth of incorrect strategic assumptions. What he and his team should have done, Ries writes, was only what was necessary to illuminate what customers value. By learning that, the organization could change its strategy and the scope of projects to something that will drive real success—not just on-time, on-budget and within-scope projects. He writes:

> I've come to believe that learning is the essential unit of progress for startups. The effort that is not absolutely necessary for learning what customers want can be eliminated. I call this validated learning. ... As we've seen, it's easy to kid yourself about what you think customers want. It's also easy to learn things that are completely irrelevant.

Think about how this definition of progress overturns old ideas familiar to project managers and their teams. Progress has traditionally meant moving from one project phase to another while remaining on schedule and within budget. Progress can and should occur, in this old framework, by simply executing the plan approved by the organization.

In contrast, Ries argues that teams must avoid wasting their time and the company's money by learning what customers want as soon as possible. (Systematically eliminating wasteful processes is at the heart of "lean" manufacturing techniques pioneered by the Toyota Motor Corp.; Ries's book adapts this concept to the startup context.) If that means throwing out the original plan—and forcing the original project to "fail"—so be it. The sooner that failure occurs, the better.

And the sooner organizations can learn to embrace these kinds of failures—the kind that illuminate flawed assumptions about how customers work and the world works—the better. This sort of shift to embrace learning through failure, to build a culture of learning, isn't easy. At most large organizations, there is a deeply embedded culture reinforcing old conceptions of "success." Leaders have effectively trained project managers and PMOs to only execute solutions to the problems they have defined. But today, as the environment companies compete in changes rapidly, PMO and project leaders must always be mindful of whether they're solving the right problems. If discovering the company is focused on the wrong problem, or is imagining a problem that doesn't exist, means "failing fast," then so be it. Such failures should be rewarded for not only avoiding wasted time, energy and money but for developing organizational intelligence—for teaching the organization how it must adapt itself to the world as it really is.

NASA hosts "My Best Mistake" website a collection of stories told by project managers and knowledge practitioners in the NASA Community. Each story tells how the author learned a lasting lesson from a mistake. (https://km.nasa.gov/tag/my-best-mistake/)

As painful as failure can be, it is often how innovations occur. We all love the idea of creative bolts of lightning striking genius inventors, or of a brilliant CEO like Steve Jobs pointing his teams in the right direction from day one. Reality is far more complicated and painstaking: Jobs was more like an editor, rejecting and refining ideas and designs until Apple had arrived at true innovations. Referring to SpaceX, the space transport company he founded, Elon Musk once said, "Failure is an option here. If things are not failing, you are not innovating enough."

If an organization can't learn to fail quickly and thoughtfully, it will fail to learn. And if it fails to learn today, then someday in the future some big lessons will be forced upon it—to the detriment of its bottom line. It is also important to point out that the "failing fast" mantra has become fashionable, and people get carried away, without truly understanding it. As Marc Andreesen, a Silicon Valley entrepreneur, investor, and software engineer remarked in a podcast interview, "It is not about failing fast but succeeding in the long run. They are not the same thing."

One way to succeed in the long run is by making failure survivable.

HOW TO MAKE FAILURE SURVIVABLE

If you learn something from a failure, it's not truly a failure.

In *Adapt: Why Success Always Starts with Failure*, author Tim Harford pulls back the curtain on how success in the business world really works. Major companies do not rise like a rocket to take over the world. In fact, although "few company bosses would care to admit it," Harford writes:

> The market fumbles its way to success, as successful ideas take off and less successful ones die out. When we see the survivors of this process—such as Exxon, General Electric, and Procter & Gamble—we shouldn't merely see success. We should also see the long, tangled history of failure, of all of the companies and all of the ideas that didn't make it.

Harford explores an idea that resonates and further elaborates one of the central themes of this book of a shift from a mechanical to an organic approach. Modern organizations have much in common with the biological process of evolution. In that process, the fittest variations of a species pass on their genes to the next generation, while the unfit die. In the realm of business, we tend to imagine brilliant leaders as pushing grand successes, but in fact, success usually emerges gradually, in fits and starts. Failure is a *normal* part of the process of creating improvements. Harford writes:

> Selection happens through heredity: successful creatures reproduce before they die and have offspring that share some or all of their genes. In a market economy, variation and selection are also at work. New ideas are created by scientists and engineers, meticulous middle managers in large corporations or daring entrepreneurs. Failures are culled because bad ideas do not survive long in the market place: to succeed, you have to make a product that customers wish to buy. ... Many ideas fail these tests, and if they are not shut down by management they will eventually be shut down by a bankruptcy court. Good ideas spread. ... With these elements of variation and selection in place, the stage is set for an evolutionary process; or, to put it more crudely, solving problems through trial and error.

PMOs can be a crucial part of this "trial and error" process— they should promote adaptive planning and execution based on lean and agile ideas of experimentation to put out minimum viable products (MVPs) to seek rapid customer feedback, learn, and adjust. PMOs should also be part of the self-correcting process by which failing ideas are "shut down." The PMO can play a much bigger role than simply flagging failures to executive sponsors and other senior leaders to ensure that fatally flawed projects are put out of their misery as quickly as possible.

Based on data and analysis, the PMO can point out that success just isn't in the cards for some projects. It's not an easy thing to do. As we have discussed in previous chapters, it is hard to make tough decisions and kill projects in flight. But healthy organizations build self-criticism and adaptive mechanisms into their DNA to avoid becoming a lumbering giant that blindly walks off a cliff. With its unique inside-out position—it has insider project knowledge, while still being outside the team and its sponsoring department—a strong and valuable PMO should have the authority and independence to call out hopelessly off-course projects when it sees them.

PMOs can add real value by skillfully helping teams, and organizations more broadly, learn from the failure. This means going beyond the normal lessons learned process of documentation. It means reminding teams that a "failed project" isn't a failure if the organization learns something. It means highlighting lessons learned from a failure, or highlighting a capability that came out of a failed project.

This amounts to nothing less than helping project managers and their teams reconceive what "success" is and what the goal of projects is. If a PMO can enable learning from failures to help teams across the organization improve, so much the better.

Traditionally, PMOs (and project managers) are risk-averse: They view risks as threats. The scariest risk of all, of course, is that the entire project is worthless to customers: they won't flock to a new product or service, or they'll have no use for new features. But if a project's projected benefits are useless to the customer, then everyone will be better off realizing this risk as quickly as possible: the customer won't feel misunderstood and insulted, the project team won't feel demoralized after a stretch of hard work, and the organization will avoid wasting money.

The PMO, then, should work to help teams understand that it is a good thing to realize these kinds of risks because they offer crucial chances to learn. All risks should be reconceived as opportunities, in a sense—opportunities to winnow out the bad ideas and advance toward the true innovations that will ensure long-term survival.

Beyond making failure survivable, how could you take it further and exploit it to become better and stronger from it? Nassim Taleb, in his book, *Antifragile: Things that Gain from Disorder*, explains:

> Some things benefit from shocks; they thrive and grow when exposed to volatility, randomness, disorder, and stressors and love adventure, risk, and uncertainty. Yet, in spite of the ubiquity of the phenomenon, there is no word for the exact opposite of fragile. Let us call it antifragile. Antifragility is beyond resilience or robustness. The resilient resists shocks and stays the same; the antifragile gets better.

Without realizing our methods and approaches are designed to make our environment, and projects weaker, we cannot withstand the DANCE and impact of risks and variances. Project managers and PMOs should question how you can design your project environment to add stressors and variability so that the teams can learn and grow stronger from it.

THE DNA STRANDS OF LEARN: 7 C'S OF CULTIVATING A LEARNING ENVIRONMENT

Figure 10.1 lists the 7 C's of cultivating a learning environment and PMO. Think of these learning elements as the strands that enable the DNA to evolve and mutate based on changing conditions.

Culture

Making failure acceptable and learning from it is easier said than done; it is a cultural issue, and it is going to need a mind-shift to foster and cultivate the right conditions for it. We have discussed tips of how project managers and PMOs can get started to learn from failure and make failure survivable. Another dimension is how to foster a learning environment and culture, where there is a desire, passion, and excitement for learning.

In her book *Fierce Conversations,* Susan Scott writes: "Burnout doesn't happen because we're solving problems, it occurs because we're trying to solve the same problem over and over."

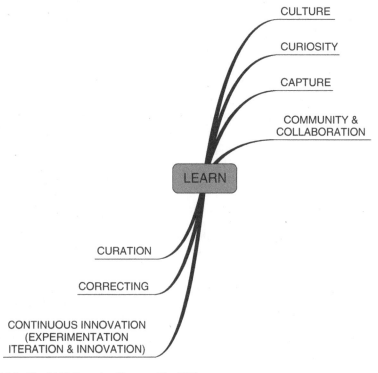

Figure 10.1: The DNA Strands of Learn—The 7 C's

Project managers know that every project is unique, but still feel like they are solving the same problem over and over. What if project managers and PMO could foster a culture where every project is viewed as a unique learning opportunity, and they treat execution as learning? Moving beyond execution to studying how to make it better. In a *Rotman Magazine* interview, Amy Edmondson explained why it's so important to adopt a mindset of "execution as learning":

> *Execution as learning* means getting work done while being highly engaged in the process of finding ways to do it better. Its defining attribute is the integration of constant, sometimes unremarkable, small-scale learning into day-to-day work. ... In these cases, workers see the work as an interesting puzzle at all times.

> We tend to think of learning and execution as separate entities: you get your work done, and you do your learning offline. ... But that isn't how it works anymore in a fast-paced field of any kind: you need to be *constantly* learning.

If we all approached projects as interesting puzzles, as interesting learning experiences, then "solving" them can become more engaging. And setbacks, or outright failure, won't be as personal and painful. Instead, they'll be opportunities to gain insights into what the customer wants or shortcomings in organizational capabilities.

Of course, a learning culture has to involve more than simply changing how we perceive projects and their execution. They have to create the right atmosphere using gamification, competition, and contests to create an environment where learning is cool, fun, and rewarded.

Another element of a learning culture that is overlooked is to create an environment where it is OK to slow down, question, and reflect. When project managers spend their days running from one fire to another, without any time to stop and think, to step back and assess whether a project no longer makes strategic sense or has some other fatal flaw, a culture of learning isn't likely to flower.

Rather, the ability to learn grows out of dedicated time for reflection and mindfulness for introspection and after-action reviews. Too often, the "lessons learned" process isn't successful because people don't (or can't) stop and think to take it seriously. "Learning" becomes a rote exercise before team members head off to another project, never to look backward as we saw with Cathy's scenario above.

Contrast this with the company contracted to restore the Acropolis of Athens, upon which the Parthenon and other ruins from the fifth century B.C. stands in Greece. The leader of that project scheduled regular times for work stoppages, allowing team members to reflect and think carefully about any mistakes made.

Building this kind of practice into project execution is how a culture of learning can truly develop within an organization. Ideally, individuals are asking questions without fear of embarrassment or reprisal, sharing information, seeking help, experimenting (within reasonable boundaries), and seeking feedback. All of this is what learning is made from. We shouldn't only celebrate project successes—we should celebrate learning in all its forms as a sign of organizational health and progress.

To cultivate a learning culture, explore ways to promote execution as learning, make learning fun and rewarding, and inspire questioning and reflection.

Curiosity

"I have no special talents. I am only passionately curious."

Albert Einstein

Traditional cultures suppress curiosity and questioning where you are supposed to stay within the boundaries. In today's DANCE-world that is in a constant state of flux, curiosity is essential to remove the blinders. Project and PMO leaders need to spark an insatiable sense of curiosity and questioning to deal with increasing ambiguity and uncertainty.

They need to normalize and encourage curiosity. In his book, *A More Beautiful Question: The Power of Inquiry to Spark Breakthrough Ideas,* Warren Berger reveals how highly innovative businesses are fueled by employees' ability to ask fundamental, game-changing questions. Noting that Google is at the leading edge of how companies are trying to transform the workplace into a "learn-place," Berger writes:

> The company established Google University as a platform for bringing in guest lecturers, then went a step further in creating Googler to Googler, a program in which Google employees host in-house classes to teach other Google employees. Not surprisingly, there are courses on technical or business skills, but the curriculum also includes courses on public speaking and parenting. Former Google engineer Chade-Meng Tan even teaches a course on mindfulness (useful in helping one to step back and question).

Part of cultivating a hunger for learning and curiosity among employees involves shifting the organization's culture away from an expert mindset. This is not easy; to prove their value, many people feel the need to assert their expert knowledge in front of colleagues (or at least their boss). Project managers are as guilty of this tendency as anyone else, along with PMO leaders. It's a natural and understandable instinct, especially in a competitive environment.

But if everyone believes they already know all that they need to know to succeed—well, you can see how vulnerable this would make any organization to change. As Epictetus said, "It is impossible to begin to learn that which one thinks one already knows." This is why humility is such a valuable quality: it opens up the space for curiosity and helps people shift away from an expert mindset to become comfortable in a learning mindset.

Ultimately, when curiosity is in the air, and a normalized and palpable aspect of an organization's culture, complacency will be less common. There will be a degree of comfort with ambiguity and dealing with the unknown. More questions will be asked, and answers won't be defensive. They will be willing to try new things. It is important to drive toward applied curiosity. Only application of curiosity will open new doors and drive innovation.

Capture

Learning is an inherently cumulative act. We build on knowledge we already have to develop more sophisticated ideas, more advanced skills, and wisdom. Learning is impossible with a functioning memory bank.

So imagine an organization that has no memory bank, no ability to remember what went right and what went wrong. It would fail quickly. Conversely, organizations with robust mechanisms for capturing and disseminating lessons learned learn from mistakes, avoid repeating failures, and recalibrate their activities toward higher-value goals that align with the operating environment.

At least that's the theory. Knowledge management, document repositories, and collaboration tools to capture project artifacts are a foundational aspect, but not enough. Although the tools have gotten much better, the challenge always has been how to capture the tacit knowledge and insights that are hard to codify in a database. As I noted earlier, the traditional approach to lessons learned—capture them at the end of a project and file them away for other teams to digest before embarking on a similar project—rarely works in practice. It's also inadequate: Why wait until the end of a project to think seriously about what went wrong?

Following are ways that next-generation project and PMO leaders can elevate their lessons learned approach to effectively harness and apply lessons learned.

Feedback Loops

First, feedback from customers and stakeholders needs to be captured and tracked in an ongoing fashion. At every opportunity, *feedback loops* should be built into the project lifecycle. Solid stakeholder management practices are just the starting point; feedback elicited from stakeholders (especially the customer) needs to be immediately distributed to the team as necessary so project plans can be appropriately recalibrated. This sort of course correcting should be constant.

But there are less orthodox approaches available beyond this kind of positive feedback loop. And they stem from the recognition that feedback loops are often inadequate in highly disruptive industries. The whole idea of feedback is to learn from experience and analyze actual performance in relation to planned performance. But when the future isn't necessarily linked to the past, feedback approaches can come up short.

That's why it is important to distinguish between single-loop and double-loop learning. Academic Chris Argyris coined the term *double-loop learning* in his book, *Teaching Smart People How to Learn*. He explains the difference using this analogy: A thermostat that automatically turns on the heat whenever the temperature in a room drops below 68°F is a good example of single-loop learning. A thermostat that could ask, "Why am I set to 68°F?" and then explore whether some other temperature might more economically achieve the goal of heating the room would be engaged in double-loop learning.

The repeated attempt at the same problem, with no variation of the method and without ever questioning the goal, as opposed to changing the mental model on which a decision depends is double-loop learning.

In Chapter 8, I discussed the example of Kumar, who seems to be caught in a never-ending loop of fixing bugs, without questioning why the bugs are occurring in the first place. He was getting feedback, which kept him in the recurring loop. Designing the right feedback approach is critical to learn and evolve. Also, Chapter 8 described how metrics and measurements are mechanisms for feedback.

Feed-forward

Feed-forward approaches address a temporal disconnect by guiding leaders to make choices by working backward from an anticipated future. The most well-known feed-forward technique is scenario building, wherein different future scenarios for an organization are around possible futures with varying kinds and degrees of uncertainty. The appeal of this approach to any team tasked with charting an organization's strategy is obvious, but the technique can also be leveraged by PMOs engaging project teams, or a project leader engaging his or her team. How could the organization's context shift during the course of the project in ways that undermine its promised benefits? How might the assumptions driving the project's business case be proven wrong? The idea is to help people be comfortable with asking challenging questions in the ever-changing environment.

Retrospectives

Retrospectives are an approach that is used in agile methodologies to reflect on how to become more effective and adjust behavior and processes accordingly. Retrospectives are designed to primarily address three questions at the end of each iteration—what worked well? What did not work well? What actions can we take to improve our process going forward? A key principle of making the retrospective work is to separate the people from the process. The emphasis is on the principle that "regardless of what we discover, we understand and truly believe that everyone did the best job they could, given what they knew, at the time, their skills and abilities, the resources available, and the situation at hand."

There are many variations of making retrospectives more engaging, interactive, and meaningful. Starfish is one example where you sketch a starfish on a flipchart with five sections—keep doing, less of, more of, start doing, and stop doing—and the team reflects and writes down ideas on stickies and sticks them in appropriate sections on the chart. Another one is to write the acronym PANCAKE on the flipchart, which stands for what puzzles, appreciation, news, challenges, appreciation, knowledge, and endorsements, and the team reflects on these as a part of the retrospective.

Another variation of the retrospective is a futurespective that flips the focus from reflection to prediction. It is a virtual journey into the future to question what would we need to accomplish our goals.

Pre-mortem

Pre-mortems also flip the normal script. While retrospectives are held after the iteration or lessons learned are conducted after the project is closed, they are more like post-mortems. Pre-mortems are conducted before the project has begun—and assume it has failed. Matthew Syed describes it in his book *Black Box Thinking*:

> With this method, a team is invited to consider why a plan has gone wrong before it has even been put into action. It is the ultimate "fail fast" technique. The idea is to encourage people to be open about their concerns, rather than hiding them out of fear of sounding negative.

The pre-mortem is crucially different from considering what might go wrong. With a pre-mortem, the team is told, in effect, that "the patient is dead": the project has failed; the objectives have not been met; the plans have bombed. Team members are then asked to generate plausible reasons why. By making the failure concrete rather than abstract, it alters the way the mind thinks about the problem.

According to celebrated psychologist Gary Klein, "prospective hindsight," as it is called, increases the ability of people to correctly identify reasons for future outcomes by 30 percent.

Storytelling

Lessons learned documents tend to be full of bullet-point lists of tips and links to templates and tools—and teams searching for wisdom often encounter similar lessons across many different projects. In other words, the lessons learned can feel generic and therefore useless, which means that over time teams will stop bothering to even look through documents for anything pertinent—and then walk blindly into avoidable project challenges.

Done right, lessons learned should provide deeper knowledge and insights. They should answer these questions:

What worked? What did not work? Why?
Who else has tried this approach before?
Who are the expert resources who are good at this kind of project?
How were the key challenges overcome?
What led to end user and customer satisfaction?
Based on the experience, what would they have done differently?
What were the pitfalls to watch for?

To be effective, avoid the bare bones of bulleted lists. Instead, dress lessons learned up in a narrative style to offer context and background. Organizations like NASA, the World Bank, and many others use stories to gather and share project lessons learned. Project managers get to reflect, compose, and share their real-life experience on the project in a natural way and in the form of a story. Valuable tacit knowledge that cannot be captured in a formal report can be harnessed in a story.

The following are examples of tips from a lessons-learned document:

- It is important to identify hidden stakeholders.
- Stakeholder management should be performed throughout the project life cycle.

Here is the story after transforming these points into a lessons-learned anecdote:

> *As we got into the crucial stage of project execution of breaking ground to dig the tunnel, we had to stop work unexpectedly because a stakeholder showed up, and who we had failed to identify in our stakeholder management process. Guess who the stakeholder was? It was a red viper snake, which is an endangered species. Due to the strict environmental laws, the project was stopped until we could relocate the snake's nest. We had the resources committed, and the delays cost the project over US$225,000, pushing the schedule back by a week.*

Effective stories have a high learning value. They do not have to be long or be high drama—they can be simple and to the point. Generally, lessons-learned stories should focus on the mistakes made and show how they were corrected, with an explanation about the alternative and why it worked.

There's a role for the PMO to play here: it can host a storytelling session instead of formal post-project reviews and a lessons-learned session. The ultimate goal of any lessons learned process is simple: get people to read and remember them. That is how organizations avoid repeating failures. So instead of *lessons lost* by having projects continuously fall into the same holes, use stories to harness *lessons learned* and transform them into *lessons applied*.

The Learning Question

Any of the lessons learned, storytelling or coaching should include what Michael Bungay Stanier deems the learning questions, "What was most useful to you?" In his book, *The Coaching Habit,* he explains that people don't even really learn when they do something. They start learning, start creating new neural pathways, only when they have a chance to recall and reflect on what just happened, and the learning question helps prompt that.

Community and Collaboration

Learning—at least in the context of an organization conducting team-based work—is inherently social. Building a culture of learning and curiosity, and encouraging the capture, and sharing of lessons learned: All of this is only possible if employees feel they are part of a meaningful community. As we discussed in Chapter 7, the PMO can be instrumental in building a sense of community. Instead of the PMO being at the helm and disseminating top-down prescriptions and the right way of doing things, it becomes the facilitator. It recognizes the effectiveness of informal structures to promote learning and sharing of knowledge.

If you think about it, this way of harnessing and disseminating institutional knowledge and lessons learned is healthier and more sustainable that the traditional PMO-as-expert mode. In some ways, it's harder because the PMO's role isn't as formalized. But it can be a powerful paradigm for connecting diverse people in meaningful ways that foster knowledge sharing and the real learning of lessons. You can start small by sponsoring lunch and learn sessions and grow the learning community organically with more activities and events over time.

Project teams are by definition collaborations, and inevitably they develop their own learning culture to move the project forward. As we discussed in Chapter 7, modern matrixed organizations are full of disconnects. In global organizations employing tens of thousands of people, people find themselves seemingly trapped in a silo and cut off from the rest of the company. There are real structural obstacles facing anyone who wants to connect someone with certain expertise to explore potential collaborations. Think of all the wasted energy and talent, all the missed opportunities for learning and knowledge advancement in such compartmentalized organizations.

The PMO can work to connect people in ways that spark collaborations. It can implement tools that automatically connect individuals to relevant experts across the organization—even if they're outside their network of contacts or business division. Knowing who to ask in a large organization is no simple task. It may involve long searches and frustrating dead ends. This is where collaboration tools like Starmind, which uses artificial intelligence to automatically route any employee's question to the right expert within your organization can be effective. New hires, seasoned experts with knowledge gaps, shy employees—all get their answers from the best source.

It allows employees to easily answer questions posed by employees anywhere in the world and promote learning in real time across borders, sharing knowledge and best practices. Among Starmind's global clients, employees in another country answer 70 percent of their colleagues' questions. Spanish telecommunications company Telefonica is one of those clients. It wanted to build a culture of collaboration, so it fully integrated Starmind into its employee portal under the name "DigitalBrain." The result, according to Starmind, is that expertise is now available to everyone, everywhere, all the time. Expertise moves "freely and neutrally through the organization. Old inherited structural barriers are erased."

Next-generation collaboration platforms break down obstacles to learning, connecting people across silos and stimulating learning that can prevent duplicative projects and unlock creativity and innovation.

Curation

> "Curation is the ultimate method of transforming noise into meaning."
>
> *Rohit Bhargava*, Non-Obvious 2017

We live an age of information overabundance. Every day, we're bombarded with information from a growing array of devices and applications connecting us to more and more, and we don't even have time to look up from our devices. One role that has emerged is the role of curators in many fields to make sense of all the noise into meaningful choices. This is a great opportunity for the next-generation PMO to be the curator—to identify, organize, and share lessons, ideas, best practices, tools, and apps. They can curate as well as actively search for the best in class in each of these areas that might be useful to the community.

As a curator, the PMO ought to be regarded as the essential go-to resource for all things related to project management and strategy execution. Museum curators do more than just create exhibits; they also work to grow the institution's permanent collection of art. In the same way, a PMO should do more than just hold up best practices and lessons learned to whoever knocks on its door for help. Its leaders should fan out across the organization hunting for new approaches to strategy execution, innovation, knowledge sharing, and collaboration.

Correcting (Self-Correction and Application)

There is a tendency to overestimate the role of planning beforehand, and underestimate the role of correction, after kick-off. As Rolf Dobelli in his book *The Art of Good Life* points out, "As an amateur pilot I've learned that it's not so much the beginning that matters but the art of correction following take-off. After billions of years, nature knows it too. As cells divide, copying errors are perpetually being made in the genetic material, so in every cell, there are molecules retroactively correcting these errors. Without the process of DNA repair, as it's known we'd die of cancer hours after conception. Our immune system follows the same principle. There's no master plan because threats are impossible to predict. Hostile viruses and bacteria are constantly mutating, and our defenses can only function through perpetual correction."

Constant course correction and readjustment after take-off is critical, but there is another important lesson we can learn from aviation. Statistically, flying in a plane is safer than driving a car. How did it come to pass that it is safer to rise tens of thousands of feet above the ground than to drive down the street? The short answer: the application of lessons learned. Aviation regulators and the airline industry created incredibly robust processes that pilots and crew members must follow before takeoff. After each flight, they document what happened. All details, including any malfunctions and how airline employees responded to them, are put in a massive anonymous database that is accessible to all airlines. Anyone can see anything—the flow of knowledge is unimpeded between airlines and their pilots. Think about how you can emulate this in your environment and promote radical capturing, sharing, and transparency across projects. Not just sharing, but cultivating a culture that it's OK to fail, as long as we can learn, apply, and evolve from it.

At Google, there's a rule: It's OK to fail, but not in the same way twice. In other words, you're expected to learn from mistakes. The challenge is that you may address some of the issues with lessons learned discussed above and do a decent job of capturing lessons learned, but they are of no use if they are not applied. Often people are demotivated about lessons learned because there is an underlying perception that it is a fruitless exercise. They are demotivated because the belief is "What's the point, we already know the lessons ... but we can't do anything; the organization is not set up or ready to apply them."

To apply the lessons and help prevent repeat failures the organization or the PMO has to ask the right questions to uncover what went wrong and what can be done to change processes to avoid mistakes or build self-correcting mechanisms. In his book *Seeking Wisdom*, Peter Bevelin explains how to ask the right questions to avoid repeat failures:

> How can we create the best conditions to avoid mistakes? How can we prevent causes that can't be eliminated? How can we limit the consequences of what we want to avoid? How can we limit the probability of what we want to avoid?

The goal is to methodically interrogate the failure to understand what contributed to it and why, and how those factors can be eliminated or avoided. Table 10.2 offers a series of questions to ask, depending on what someone is trying to avoid repeating.

Table 10.2: How to Avoid Repeat Failures

What to Avoid	Cause	Antidote
What were the mistakes?	Why did those happen?	What are the major risk factors? How do specific errors evolve? What factors contribute?
Stupidity/irrationality	Big idea that helps explain and predict?	What is rational? How can I create the best conditions to make good decisions? What can be eliminated or prevented?

Source: Adapted from Peter Bevelin (2007), *Seeking Wisdom.*

Checklists

One tool for preventing repeat failures and applying lessons learned is simple and familiar: the checklist. Overtime, the learnings can be harnessed to build and improve checklists for repeatable processes. As pointed in Chapter 9, Harvard Medical School professor Atul Gawande makes the case that checklists can help avoid repeat failures.

In highly specialized fields such as medicine and aviation, it doesn't matter how well trained and intelligent a surgeon or a pilot is—inevitably, something will fall through the cracks because the systems to be managed and procedures to be executed are so complicated. This is why airlines embed checklists in their day-to-day flight operations: to prevent inevitable human error.

In his book, Gawande notes that a five-point checklist implemented by Johns Hopkins Hospital in 2001 nearly eradicated central line catheter infections in the intensive care unit, preventing more than 40 infections and 8 deaths over a period of 27 months. Similarly he provides examples of successful use of checklists in skyscraper construction projects.

The possibilities are limitless—checklists could potentially incorporate items to stave off all the failures that have bedeviled organizations in prior years. If team members can recognize their fallibility and see checklists as powerfully simple tools for preventing repeat mistakes, progress is assured.

CONTINUOUS INNOVATION

Moving from Continuous Improvement to Continuous Experimentation, Iteration, and Innovation

The traditional notions of learning and maturity focus on continuous improvement, which is important but not enough. The assumption is that you are improving the same thing over and over to gain efficiencies and get better. In a constantly changing and disruptive world, continuous improvement is like running better and faster, just to stay in the same place, whereas continuous innovation is a double loop, where you learn and evolve to create something new and better.

To get ahead, you have to evolve from continuous improvement to continuous innovation. You take the outputs you generate, and turn the data and learning from it, as inputs for innovation. As Shane Parrish, an influential blogger and founder of Farnam Street Media, discusses in an interview, "The outputs of Amazon are now becoming the inputs, and that's how they just keep getting better and better. They are quick to adjust, and that's part of the reason that they are so dominant." Another point to emphasize is that the learnings have to also include an outside-in customer perspective. As Shane continues, "They have a 'customer first' mindset that allows you not to have an ego. I think the feedback loop model is much better, where you remove people's egos and focus on the customer instead."

A PMO's job is never done—and its leaders should relish this fact. New employees will always be coming and going; new teams will be assembled and disbanded, new lessons learned will be collected and distributed. All the while, the organization's external context is changing, along with—hopefully—its strategy and project portfolio. Ultimately, the value of that portfolio derives from the amount of innovation it contains.

Traditionally, at a basic level, the PMO has existed to help an organization hold itself to high standards. When looked at from the C-suite, it's a complex, multifaceted quality control mechanism. But in today's competitive landscape, the PMO must be more than just a purveyor of best practices that allow the organization to improve project delivery. It must go beyond mere incremental improvement to drive continuous innovation with a customer-first orientation.

Meta-Learning

One person who has spent much of his career dedicated to the study of learning in many different forms is the author, entrepreneur, self-proclaimed "human guinea pig," and popular podcaster Tim Ferris. In one of his books, *The 4-Hour Chef,* he distills a meta-learning technique for learning any skill, which should resonate with project and program managers. The four-step process is simplified by the acronym DiSSS:

Deconstruction. "What are the minimal learnable units, the LEGO blocks, I should be starting with?"
Selection. "Which 20 percent of the blocks should I focus on for 80 percent or more of the outcome I want?"
Sequencing. "In what order should I learn the blocks?"
Stakes. "How do I set up stakes, to create real consequences, and guarantee I follow the program?"

Why this is particularly important for project management is because project managers are trained and supposed to be good at the first step of deconstruction and breaking things down. They can leverage the other three steps of selection, sequencing, and stakes in not just learning but also effectively planning the project or program. By using these steps, they can quickly learn to maximize outcomes, by selecting the right tasks and sequencing them in an effective way, while focusing on the right stakes and stakeholders.

> "In times of change learners inherit the earth; while the learned find themselves beautifully equipped to deal with a world that no longer exists."
>
> *Eric Hoffer, author, social philosopher*

DEVELOPING LEARNING INTELLIGENCE

Use the following questions to reflect and explore ways to develop learning intelligence:

- Are we more organized to execute or organized to learn? How can we shift and balance more toward learning?
- How can we change the perception and use failure as a learning opportunity?
- How can we cultivate a learning culture?
- How can we build better feedback loops?
- How can we encourage teams to speak up if a project is doomed to fail?
- How can we adapt lean and agile approaches for reflections and lessons learned?
- How can we get experts to share?
- How can we reward sharing and learning?
- How can we become the curator of ideas, best practices, and lessons learned to create meaning from noise?
- How can we improve the taxonomy of knowledge management systems?
- How can we better utilize collaboration platforms?
- How can we inspire and promote curiosity to create an infectious learning environment?
- How can we make failure survivable and build course correction in project management processes?
- How can we create the best conditions to avoid mistakes?
- How can we prevent causes that can't be eliminated?
- How can we limit the consequences of what we want to avoid?
- How can we limit the probability of what we want to avoid?
- How can we better organize the study of errors?
- How can we apply metrics to measure learning as a unit of progress?
- How can we build community and collaboration to promote sharing and learning?
- How can we move from continuous improvement to continuous innovation?
- How can we leverage and improve meta-learning and accelerated learning techniques?

KEY TAKEAWAYS

- Project management and PMOs are setup and organized to execute, not necessarily to learn. There is a difference between the two. To cultivate the right conditions for learning requires experimentation and trial and error, which is the opposite of follow the rules, do it right the first time, drive out variance, and failure is not an option.

■ Failure—or to be precise, the ability to learn from failure—is an essential part of success in today's turbulent and fast-changing business world.

■ Making failure acceptable and learning from it is easier said than done; it is a cultural issue, and it is going to need a mind shift to foster and cultivate the right conditions for it.

■ Build feedback loops and double-loop learning to ensure that you are not stuck in a repetitive attempt at the same problem, with no variation of the method and without ever questioning the goal, as opposed to questioning and changing the mental model to learn and evolve.

■ Project managers and PMOs should promote adaptive planning and execution based on lean and agile ideas of experimentation and trial and error, to seek rapid customer feedback, learn, and adjust.

■ Project managers and PMOs should aim to make failure survivable, with a focus on course correction and avoiding mistakes.

■ To develop learning intelligence, the organization must develop and promote the elements of learn—the 7C's: culture, curiosity, capture, community, collaboration, curation, and continuous innovation.

■ Go beyond traditional lessons learned to adapt agile approaches like retrospectives, futurespectives, pre-mortems, storytelling, and other techniques to better harness and share lessons learned.

■ Don't forget to have a customer-first and customer-success orientation to learn and evolve from the outside-in.

■ Continuous improvement is not enough; you must strive for continuous innovation. The PMO's highest calling ought to be to cultivate learning, to adapt, evolve, and innovate, for that is what is necessary to survive today, beyond flawless execution processes.

11

SIMPLIFY: BUILDING THE DEPARTMENT OF SIMPLICITY

"Our enemy is not the competition; it is unnecessary complexity in our processes."

CEO of a Global Conglomerate

"Perfection is achieved not when there is nothing more to add, but when there is nothing left to take away."

Antoine de Saint Exupery

Leading Questions

- Why is simplicity elegant and an essential element of the DNA?
- Why is simplicity hard?
- What are the principles of simplicity?
- How do we build a Department of Simplicity?
- How do we focus on experience and create viral customers and stakeholders?
- How do we develop simplify intelligence?

"The problem is one of perception. Most of the processes do not take that long to do," David tried to explain, "those that do are intended to provide information required to think through the project before committing resources or changing course." Ironically, the stakeholders of his PMO thought otherwise, and a few months after, I was not surprised to hear that his organization had reduced the role and nearly disbanded the PMO.

What comes to mind when you hear the term *PMO?* If you think of more work, documentation, processes, and red tape, you are not alone. In a survey by the Projectize Group, 78 percent perceived their PMOs as bureaucratic. The American Heritage® Dictionary *of the English Language* defines bureaucracy as "an administrative system in which the need or inclination to follow rigid or complex procedures impedes effective action." Bureaucracy is a critical issue for PMOs and a number one reason for push-back and lack of buy-in and acceptance for many PMOs. Unnecessary complicatedness creeps up on well-meaning PMOs like crabgrass and weeds. It is not just a PMO issue; it is an organizational challenge. Unclear, complicated processes are costly in terms of reduced compliance, rework, and frustration.

According to Yves Morieux of the Boston Consulting Group (BCG), since 1955 business complexity, as measured by the number of requirements companies have to satisfy, has risen steadily. To address each new requirement, companies typically set up a dedicated function and then create systems to coordinate it with other functions. That explains why complicatedness (number of procedures, vertical layers, interface structures, governance bodies, and decision approvals) has seen an even sharper increase. This complicatedness hurts productivity and employee engagement. Managers spend more than 40 percent of their time writing reports and between 30 percent and 60 percent of their total work hours in coordination meetings or, work on work. There is a rallying cry to simplify organizations and government, and simplification has become a strategic imperative for speed and agility in many organizations.

As Peter Drucker put it, "Most of what we call management consists of making it difficult for people to get their work done." Ironically, many projects and PMOs reflect a similar reality. One of the most common complaints about project management and PMOs is that they tend to make things more complicated than necessary. It is not just a perception problem; it is also the reality. For example, many PMOs will require even simple projects to follow an excruciatingly detailed methodology or file a monthly report that takes longer than a month to produce! If you don't follow the PMO process your project plan is red-lined, you might get audited, and you will have to provide additional documentation. The shortcoming of the PMO is not in what it does but what it overdoes.

KILL THE PMO?

"When will the PMO stop us from conducting business?" was a comment overheard from a frustrated executive in a financial services organization. Often, PMOs are guilty of unclear, complicated processes that are costly in terms of time, rework, frustration, and simply conducting business. These complicated processes are like creepers and weeds that can spread and strangle healthy plants and trees if not controlled in time.

The PMO needs to kill the traditional perception of bureaucracy and reinvent itself as the Department of Simplicity. You can take a proactive approach like an information technology (IT) PMO, which conducted a bonfire of their old processes and methodology and invited stakeholders to the reinvented *Department of Simplicity!* The vision for starting and sustaining PMOs should be that they are the *Department of Simplicity* within the organization. This presents an opportunity for the PMO to be the agent of simplicity and drive simplification and dedicate itself to identifying and reducing unnecessary overhead and complicated processes.

WHY IS SIMPLICITY A STRATEGIC IMPERATIVE IN TODAY'S DISRUPTIVE DANCE-WORLD?

To survive in a disruptive DANCE-world, speed and agility are essential and simplicity is a strategic imperative. How do you create a start-up culture and entrepreneurial spirit and guard against bloat and bureaucracy that slows you down? As one executive in a global conglomerate remarked, "Our enemy is not the competition; it is unnecessary complexity in our processes." You have to create a culture where you can work together and focus on initiatives and projects that matter the most, make jobs easier, simplify processes, and enhance customer experience. Organizations and big established companies like ConAgra, General Electric, Phillips, Vanguard, and others have embraced simplification as a strategic imperative and have different aspects of corporate lean, fast-track, and simplification initiatives.

"Execution travels at the speed of sense-making," according to Bill Jensen, known as Mr. Simplicity, who has been writing and extolling the virtues of simplicity for the past 25 years. If the PMO can simplify and create less clutter and more clarity, and help everyone make sense of it faster, it can catalyze strategy execution.

We all face a cognitive overload of choices as we are bombarded with information and stimuli from multiple sources. The way to connect and grab attention from competing channels is to cut through the clutter with simplicity and elegance that can surprise and delight the PMO customers and stakeholders in an increasingly complicated world.

Simplicity can have an impact on revenue. Siegel + Gale, a strategic branding firm, has created a global brand simplicity index, which has a portfolio of brands/companies that are perceived to offer a simpler experience. The study revealed interesting insights—75 percent of consumers are more likely to recommend a brand if it offers a simpler experience. One hundred percent of brands/companies in the simplicity portfolio have beaten the average global stock index since 2009.

Optimizing for Efficiency versus Optimizing for Simplicity

The initiating purpose of project management and PMOs is to bring about consistency and efficiency by implementing standards and processes. However over time they proliferate and create unnecessary "complicatedness" in the organization. For example, a new stage-gate process, while bringing discipline to development, was also bringing additional reviews and paperwork. One of the teams estimated that they were spending upwards of 70 percent of their time preparing for project reviews, attending review meetings, or responding to review issues that left little time for actual project work.

Typically, PMOs are geared toward optimizing for efficiency. As illustrated in Figure 11.1 it is important for the PMO to understand this distinction between optimizing for efficiency versus optimizing for simplicity.

Optimizing for Efficiency	Optimizing for Simplicity
Mechanical perspective	Organic perspective
Inside-out approach	Outside-in approach
Addition of structure/process/rules	Subtraction of structure/process/rules
Exploitation and adherence to structure and process	Exploration and flexibility of structure and process
Streamline	Redesign
Procedural focus	Behavioral focus
Drive out variance	Use variance to analyze & improve
Passion for process	Passion for end-users & customers
Focus on outputs & efficiency	Focus on outcomes & experience
Measure efficiency & compliance	Measure effectiveness & experience
Faster, cheaper, better from organization's perspective	Faster, better, cheaper from customer's perspective
© J. Duggal	

Figure 11.1: Optimizing for Efficiency versus Optimizing for Simplicity

When you optimize for efficiency, you think inside-out from the PMO's perspective; when you optimize for simplicity, you will think outside-in from the end-user and customer perspective. Paradoxically, efficiency focuses on adding or streamlining structure, process and rules, and yields outputs. Simplicity aims to reduce or re-invent processes based on effectiveness and experience and results in greater satisfaction and better outcomes and results. Optimization aims for faster, cheaper, and better, but the impact can be different based on whether it is from the organization's perspective or the customer's. For example, a healthcare PMO tried to optimize its portfolio process with three pipelines for project approvals, which resulted in faster project approvals and was better from the organization's perspective. However, they did not think about the impact on the business customers of the PMO, as some of the same people had to participate in three different intake meetings and review additional documentation resulting in more time and frustrating experience.

Why Is Simplicity Hard?

Often, some people get confused with simplicity and have a wrong notion about it. They think simplicity means not enough depth, too easy, or a simpleton—ignorant, foolish, or silly approach. Simplicity does not mean simplistic solutions or shortcuts, lack of functionality, or limited information. Simplicity is hard; it is on the other side of complexity. An example frequently used to illustrate this comes from Mark Twain, who received this telegram from a publisher: NEED 2-PAGE SHORT STORY TWO DAYS. Twain replied: NO CAN DO 2 PAGES TWO DAYS. CAN DO 30 PAGES 2 DAYS. NEED 30 DAYS TO DO 2 PAGES.

Simple does not mean easy; simplicity is achieved only with a deep understanding of the underlying complexity. As Albert Einstein said, "Any intelligent fool can make things bigger, more complex, and more violent. It takes a touch of genius—and a lot of courage—to move in the opposite direction." Think of all the successful products and services that we all use and couldn't do without, like Google. They look simple on the surface, but to achieve that simplicity requires an immense amount of complexity.

PM and PMO Principles of Simplicity

Simplicity is difficult to practice. Project managers and PMOs can start by understanding and applying the following principles of simplicity.

From Whose Perspective?

I frequently hear from PMO managers, "Yes, we get it; we are already simplifying." Yet when we look closer through the lens of simplicity, we find that they are indeed simplifying, but from their own perspective, and not necessarily from their customer's, or stakeholder's standpoint. As Bill Jensen points out, "Throughout all history (including in all workplaces), those in power have always defined and leveraged simplicity as simpler for them, and those with less or no power have always struggled with top-down, mandated approaches that make things more complex for them." PMOs rollout processes make it easy for them to report to management, but often make it cumbersome for project managers. It is a good idea to work backward from the needs of those doing the work as you simplify. There is only one judge of simplicity: your customers and end users. You need to put yourself in their shoes and evaluate the total experience from their viewpoint. Engage them to become part of the simplification and suggest ideas and reward them for participating.

Minimalism—Less Is More

Traditional approaches rely on the principle that to control better and establish sound governance you need heavy methods, processes, and tools built on intricate rules. Simplicity is based on the opposite principle of minimalism, and less is more—seeking out the essential and separating the value-adding activities from the non-value-added minutia that sucks up time and energy. As Hans Hoffman, a legendary artist, remarked, "The ability to simplify means to eliminate the unnecessary so that the necessary may speak." The PMO needs to be in a relentless pursuit of less but better.

Scalable

Methodology and governance structures should be scalable and adaptive based on criteria like project size, scope, complexity, and business impact. One of the common complaints from project managers is that it takes them more time to complete the project documentation than the project itself. Even though it was a simple project, they had to apply all the steps to comply with the PMO methodology. To strive for global consistency and standards, a one-size-fits-all mentality sounds good but is not practical. Projects and programs by definition are unique with different characteristics requiring diverse approaches. Scalable processes and methods can be designed to address the unique aspects of projects. A simple project may need very limited process steps versus a complex project that may require more elaborate methodology steps.

Self-Eliminating

Good processes should have a built-in mechanism for changing or eliminating the process. We can all identify processes in our organizations that have survived way beyond their desired purpose. There are processes that are in practice and institutionalized simply because they have been done for a long time and nobody has questioned them. Part of the PMO governance should be a method to conduct periodic process reviews and decide when a process or practice is no longer useful or when it needs to be updated to make it useful again.

Desire Lines

Have you ever taken a shorter unpaved route while walking to get to someplace? Think about why you desired the informal way rather than the paved path. It could be for a number of reasons; it probably seemed shorter, more efficient, or you simply preferred it due to the better scenery along the way. Desire paths can usually be found as shortcuts where constructed pathways take a circuitous route. Similarly, PMOs need to sense and observe the existing desire paths of methods and processes and adapt and reinvent PMO processes along end-user, customer, and stakeholder desire lines.

Simple Rules

It is a common misconception to think that in a complex project environment with DANCE characteristics you need more top-down processes and intricate rules. Examples from the application of complexity science illuminate the opposite. In highly complex situations and uncertain and unpredictable emergent environments like nature and living organisms, simple rules are the underlying code that can handle and resolve enormous complexity. For example, I am sure at some time you have looked at the sky and wondered how some birds could fly in a V-formation without crashing into each other. This is a common example used in complexity science to illustrate the power of simple rules in nature. The birds fly based on following three simple rules: separation—maintain distance/avoid crowding neighbors (short-range repulsion); alignment—fly in the direction of the bird in front of you/steer toward average heading of neighbors; cohesion—steer toward average position of neighbors (long-range attraction).

The idea of simple rules based on complexity science has been applied to business for some time now. Donald Sull and Kathleen Eisenhardt, in their book, *Simple Rules: How to Survive in a Complex World,* explain:

> Simple rules are shortcut strategies that save time and effort by focusing our attention and simplifying the way we process information. Simple rules work, it turns out, because they do three things very well. First, they confer the flexibility to pursue new opportunities while maintaining some consistency. Second, they can produce better decisions. When information is limited and time is short, simple rules make it fast and easy for people, organizations, and governments to make sound choices. They can even outperform complicated decision-making approaches in some situations. Finally, simple rules allow the members of a community to synchronize their activities with one another on the fly. As a result, communities can do things that would be impossible for their individual members to achieve on their own.

To be effective, simple rules have to be a handful, not too many and not too few. They have to be tailored to the organization using them. They have to be well defined for an activity or decision. And they have to provide clear guidance with the freedom to exercise judgment. The PMO can use the idea of simple rules to guide effective decision making for activities like project selection and prioritization, when to kill projects, risk assessment, stage-gate reviews, and others.

In his book, *Ten Laws of Simplicity*, John Maeda, a simplicity evangelist, design guru, and former president of the Rhode Island School of Design, outlines the following 10 laws of simplicity that can be useful as you aim to simplify your PMO:

1. **Reduce.** The simplest way to achieve simplicity is through thoughtful reduction.
2. **Organize.** Organization makes a system of many appear fewer.
3. **Time.** Savings in time feel like simplicity.
4. **Learn.** Knowledge and sensemaking makes everything clearer and simpler.
5. **Differences.** Simplicity and complexity need each other.
6. **Context.** What lies in the periphery of simplicity is definitely not peripheral.
7. **Emotion.** More emotions are better than less.
8. **Trust.** In simplicity we trust.
9. **Failure.** Some things can never be made simple.
10. **The One.** Simplicity is about subtracting the obvious, and adding the meaningful.

HOW TO BUILD A DEPARTMENT OF SIMPLICITY

Assess Your Annoyance Factor

To build your own Department of Simplicity, you have to start by looking in the mirror and asking key questions to assess your annoyance factor:

- How many PMO processes do we have? Are they too many?
- Are these processes cumbersome and annoying?

- How many of the templates/reports are more than one page long?
- How many of them have more than 10 steps?
- How many of them have more than three levels of approvals?
- How much time is spent by project managers/team members on documentation?
- How much time is spent on generating, disseminating, and reviewing reports?
- What would the PMO's customers and stakeholders love for us to eliminate or reduce?
- What would those who matter most love for the PMO to stop doing?
- What is it that the stakeholders struggle with if the PMO was killed and ceased to exist?
- Is the PMO making its stakeholder's job easier?

As you ask these questions, challenge the status quo bias—the tendency to continue doing something because we have always done it this way. For example, in a financial services company, a project review process was outdated and cumbersome. Upon questioning the PMO team, nobody could identify the origin or the reasons for the process—they were just following the process blindly because it was already in place and nobody ever questioned it.

As you review and assess your project environment or PMO, you can distinguish between value-adding and non-value-adding activities based on the Japanese principles of Kaizen. Kaizen practitioners focus on *muri, mura,* and *muda. Muri* means overload, *mura* means inconsistency, and *muda* is waste. In the pursuit of implementing simplicity, overload, inconsistency, and waste are critical factors to review.

Bill Jensen also provides some tips to simplify on his simpler work blog that are particularly relevant to PMOs. Ask your user/audience/customer just one question—**"How can we make this easier for you?"**—and you will quickly learn what simplicity looks like.

Always start with time poverty and attention deficit disorder. Yes, there are bigger, more systemic, more entrenched problems that need simplification. But if you focus on saving people time and on capturing and using their attention wisely, you will never go wrong. These are among the top two challenges that will always benefit from simplification efforts.

Always give the user more control. The more control that the user/audience/customer has over using your product or service, the simpler it will be for them.

In our work with organizations around the world, we practice the following steps in our effort to simplify project management and PMOs:

- Inventory and review current processes and methods. Pick the top three processes and see how they could be streamlined, and find ways to eliminate redundancies and simplify them.
- Think about how you can cut, slim, trim, prune, combine, and modularize existing methods and processes.
- Conduct a reverse pilot, which is the opposite of a pilot. In a pilot, you test new initiatives; in a reverse pilot, you test whether removing an activity will have any negative consequences. For example, you can stop publishing a marginal report

that takes a huge effort to produce. If nobody complains after you stop it, that means the report was unnecessary.

■ While implementing methodologies, begin with an absolutely minimum set of processes. See how you can create the least annoying and least intrusive processes. Substitute practices that match more closely with your organization. Carefully add practices that address specific organizational or project situations.

■ Ensure that your PMO's methodologies and tools are scalable—fewer steps for simple projects and more detailed steps for complex projects.

■ A "one-page-fit" should be a rule of thumb for most PMO reporting and documentation requirements.

■ Identify opportunities to streamline processes, reporting structures and approvals—see how you can reduce the layers for approvals and decision making.

■ Follow this principle: don't add until you subtract. Reduce the weight of heavy methods by subtracting non-value-adding steps before introducing value-adding processes.

■ The simplified processes should be explained and communicated clearly.

■ Engaging your stakeholders early on in the co-creation of the processes or templates will increase their support and adoption of the process.

Table 11.1 is a sample grid that the PMO can create to list the process, proposed simplification, impact of the change, and who the simplification would impact as it embarks upon the Department of Simplicity initiative. You can list all the PMO processes that have been identified as complicated, cumbersome, or annoying as possible candidates for simplification.

Simplify Checklists

As we discussed in Chapter 9, part of project and PMO governance is to provide standardized checklists. Checklists can help drive the right behaviors by providing the right information at the right time. However, often checklists are too long and complicated and do not achieve the intended results. Atul Gawande in the *Checklist Manifesto,* suggests tips for simple and effective checklists: They cannot be lengthy. The rule of thumb is no more than five to nine items; They should be timely, contextual, and depend on the situation; Focus on "killer items," steps most dangerous to skip and sometimes overlooked; The most difficult part of checklists is managing tension between brevity and effectiveness. Use simple, exact, and familiar words; Checklists ideally fit on one page, free of clutter and unnecessary colors; Test in the real world and simulate. Checklists should and can be modified to fit local procedures, processes, and language.

Create a Simplicity Advisory Board (SAB)

PMO can take a leadership role and establish a simplicity advisory board (SAB). Phillips Electronics has created a SAB made up of experts who help the company create simplified offerings such as instruction manuals that non-tech-savvy consumers can understand. It is evident from Phillips's success with recent innovative home lighting

Table 11.1: PMO Sample Grid			
Process	**Simplification**	**Impact**	**Simplifies For**
Monthly program gate review	Replace with quarterly program status reports. No need to meet.	Portfolio committee is not taking action based on monthly gate review. Overall it should be a positive impact.	Sponsors, program managers, project managers, intake team
Resource management report	Remove some calculations and columns to simplify look and feel.	Simplifying the report with a better look and feel will change perception and promote greater adoption and better experience.	Program managers, project managers, business analysts, functional managers, resource managers
Required project documents	Reduce the number of project documents.	Less time to complete required documentation; better quality of documentation; less bureaucracy.	Program managers, project managers, business analysts, team members
Closing report	Make it optional for projects under 1,000 hours.	Reduce paperwork, but may promote wrong behavior of splitting projects to avoid paperwork; reduced capability to capture historical data and lessons learned.	Program managers, project managers

products like the Hue lighting system with its packaging, intuitive design, and "cool" factor. The PMO can engage a variety of stakeholders in the SAB to co-create and garner valuable feedback and advice.

RETHINKING MATURITY: SUBTRACTION, NOT ADDITION

Figure 11.2 shows the famous series of lithographs known as the "Bull" created by Pablo Picasso around Christmas of 1945. How do you think he started—from the bottom left, adding detail in each subsequent frame, or the other way, from the top right? Conventional wisdom makes us think he started from the bottom left. But actually, he did the opposite, with each iteration he subtracted, to focus on the essential elements to get to the essence of the beast. In this series of images, Picasso visually dissects the image of a bull to discover its essential presence through a progressive analysis of its form. This idea is also used at Apple to teach designers the quest for simplicity by eliminating details to create a great work of art. Apple is one company that, with its

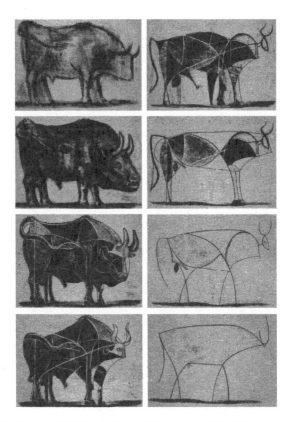

Figure 11.2: Pablo Picasso – Bull Suite of Eleven Lithographs 1945–1946
Source: By Sora [CC BY 2.0 (http://creativecommons.org/licenses/by/2.0)], via Wikimedia Commons.

products, has personified the idea that beauty and functionality come from elegant simplicity.

In the PMO context, it provides a whole new way to think about maturity. Traditionally, we have equated maturity by addition—adding more detailed standards, processes, and so on. This opens our eyes that true maturity is about subtraction and not addition. The PMO should be in the pursuit of less but better.

Creating Viral Customers & Stakeholders

Imagine instead of dreading and avoiding project and PMO activities and meetings, your stakeholders are surprised in a pleasant way. They are giving you likes and raving about the PMO and want to be part of the PMO.

We can think of similar experiences with products and services like using the iPad for the first time, visit to Disney, using websites like Google, Amazon, or eBay, or using certain apps and accomplishing what we want with few clicks in an elegant interface.

We are surprised by the simplicity and enjoy the experience. We not only want more of it, but we also want to share it with others.

When we experience a positive surprise, it compels us to do three things, according to Soren Kaplan in his book, *Leapfrogging: Harness the Power of Surprise for Business Breakthroughs:*

1. Want to experience more of it;
2. Learn about how or why it works the way it does; and
3. Share it, so that we can take a small amount of credit for others' smiles of surprise.

In this book and particularly in this chapter, I have provided several ideas to transform the experience and perception of your PMO. As the chief uncomplicator, observe and be vigilant to identify complicated processes and practices that impact customer and end-user experience. Project management and the PMO should aim to minimize the paperwork and reporting burden and ensure the greatest possible benefit and maximize the utility of information created, collected, and maintained. It should remove obstacles for the project managers and its customers and design a clutter-free, clean, and consistent PMO experience. To be effective, the PMO needs to be in the relentless pursuit of less but better. Driving toward simplicity requires persistence and vigilance, a willingness to make tough choices and ability to see the world from your customer's perspective. Simplify and make structure and processes easier from your end users' and customers' perspective. Make it fun and engaging; help them love what they hate to do. You will surprise your stakeholders with the positive experience of dealing with project management or the PMO. They will be compelled to share this experience with others and help you gain a solid following for the PMO. Can you imagine that, instead of raising their hand to kill the PMO, your stakeholders are excited to strike the famous Staples Easy button—*"That was EASY"*—as they deal with the PMO!

How will you know if the project or the PMO customers are having a delightful experience? It is important to measure and get feedback from a stakeholder perspective. The PMO can measure itself and gather valuable feedback from its customers and stakeholders by using a simple PMO delight index (PDI), which is discussed in Chapter 12.

DEVELOPING SIMPLIFY INTELLIGENCE

Develop simplify intelligence by rigorously reflecting on the following questions:

- Why are we doing it this way?
- What would happen if we stopped doing it?
- How can we reduce, cut, trim, this method or process by ___? (fill in the blank—number of steps, people, amount of time, etc.)
- How can we combine or modularize this method or process?
- Does the method or process have the right amount of scalability?

- How can we make this easier for you (customers, end users, stakeholders)?
- How can we reduce the time for this process?
- How can we reduce the reporting layers?
- How can we streamline approval?
- How can we streamline decision making?
- How can we do a better job of sense making and communicating methods and processes?
- How can we engage customers, stakeholders, and end users in the co-creation of methods and processes?
- How can we sharpen our observation and vigilance for complicatedness and opportunities for simplicity?
- How can we use design-thinking principles to design project and PMO experience?

KEY TAKEAWAYS

- There is a rallying cry to simplify organizations and government in today's disruptive DANCE-world, and simplification has become a strategic imperative for speed and agility in many organizations.
- As the PMO continues to struggle and gain buy-in and acceptance and demonstrate value, it needs to take a proactive leadership role to reinvent itself as the Department of Simplicity.
- Simplicity is not easy; it does not mean simplistic solutions, lack of functionality, or limited information. Simplicity is difficult to practice, but if project managers and PMOs adopt the simplicity principle in everything, it is sure to increase customer experience and attract a more solid following.
- To sustain and thrive, it is imperative for project managers and PMOs to constantly challenge themselves and ask key questions, like what would stakeholders want eliminated or simplified? How can we remove any obstacles for our customers and stakeholders?
- The PMOs reinvented mission should aim to minimize the paperwork and reporting burden and ensure the greatest possible benefit and maximize the utility of information created, collected, and maintained.
- Driving toward simplicity requires persistence and vigilance, a willingness to make tough choices, and ability to see the world from your customer's perspective. To be effective, project managers and PMOs need to be in the relentless pursuit of less but better.
- Reflect on the principles of simplicity—scalability, minimalism, desire lines, and simple rules—and related tools and questions to develop simplify intelligence.
- Next-generation project managers and PMOs should remove obstacles, make things take less time, and design a clutter-free, clean, and consistent PM experience that surprises delights and creates viral customers.

12

BALANCE: DANCE-ING ON THE EDGE OF CHAOS

"The test of a first-rate intelligence is the ability to hold two opposing ideas in mind at the same time and still retain the ability to function. One should, for example, be able to see that things are hopeless yet be determined to make them otherwise."

F. Scott Fitzgerald

Leading Questions

- How do we balance the opposing paradoxes?
- How do we deal with the DANCE?
- How do we thrive on the edge of chaos?
- What is the next generation?
- How do we apply and implement the DNA of strategy execution?

In the beginning of the book, we discussed how we have relied on predictive, and evolved to iterative incremental and agile approaches to deal with simple to complicated projects and PMOs based on a mechanical mindset. The question posed was how to deal with the challenges of complexity on the other side of the continuum, with increasing DANCE in a turbulent environment. Figure 12.1, Balancing along the Complexity Continuum, answers the question with a list of alternative approaches on the other side of the continuum, summarizing some of the ideas discussed in the chapters of this book.

The challenge is knowing which approach to use when, and how to balance between them. The goal should be to develop an intuitive approach based on a combination of integrative, augmented, and emergent mindset and methods.

When I was working with a group of project managers at NASA on managing complexity and dealing with the DANCE, they shared a story about a group of 23 project managers who declared mutiny midway through an advanced project

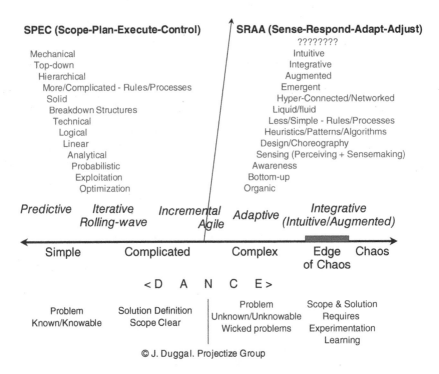

Figure 12.1: Balancing along the Complexity Continuum

management course based on NASA standard procedures. They felt the procedures were not relevant and too restrictive and they needed more flexibility to deal with the nature of their projects. The good news is that NASA's leadership realized the need to balance. An *MIT Sloan Management Review* article also describes this scenario and how they spent four months conducting interviews outside NASA. Management accepted the group's recommendations to give project managers the freedom to tailor and balance the application of standard procedures to the unique needs of their projects.

Whether you are part of a government agency like NASA, or leading a PMO at a well-established, incumbent company, or leading a project in a start-up, you have to deal with the reality of paradoxes: freedom versus flexibility; collaboration versus control; individual versus collective; quantitative versus qualitative; tangible versus intangible; data versus intuition; short-term versus long-term; outputs versus outcomes; waterfall versus agile; sustain versus disrupt; and traditional versus next generation are just some of the paradoxes that project and program managers and PMO leaders have to deal with. We have discussed these paradoxes in the chapters of this book. The paradoxes can come across as mixed messages: "focus on process, not outcomes" at the same time as "be results driven and outcome focused" can create a variety of tensions.

What creates tension can also be a source of creativity. The first thing is, instead of being paralyzed or paranoid of paradox, we have to understand its power. Albert Rothenberg, a noted researcher on the creative process, has extensively studied the use of opposites in the creative process. He identified a process he terms "Janusian thinking," a process named after Janus, a Roman God who has two faces, each looking in the opposite direction. In his research, he found that geniuses resorted to this mode of thinking quite often in the act of achieving original insights. Einstein, Mozart, Edison, Van Gogh, Pasteur, and Picasso all demonstrated this ability. Picasso's paintings reflect both calm and chaos; Einstein was able to imagine an object in motion and at rest at simultaneously.

Leaders who understand that we live in a paradoxical world and learn to manage the duality are better equipped to deal with the DANCE and disruption. They instinctively know that they don't have to focus on one or the other, but they have to balance both sides. They practice what Roger Martin calls *integrative thinking*. In his book *The Opposable Mind,* he defines it as:

> The ability to face constructively the tension of opposing ideas and, instead of choosing one at the expense of the other, generate a creative resolution of the tension in the form of a new idea that contains elements of the opposing ideas but is superior to each.

Instead of either/or, they practice but/and thinking, which can lead to new opportunities, by tapping and accentuating the positives, while tempering the undesirable side-effects.

Traditionally, project management has attracted more linear logician minds, who get perplexed when they are told, "Make sure the upcoming release is all planned and organized, and also make sure you are ready for any last-minute changes," or they cringe when they are told, "We want consistency, but be agile."

To resolve this dichotomy, project management and PMOs have to learn to apply what Professor Francesco Gino of Harvard describes as activating a "paradoxical frame." When you hear consistency and flexibility, and you view it through a paradoxical frame, you recognize the inherent incompatibility, but you also understand the potential for consistency and flexibility to positively reinforce one another. Otherwise, you might focus on only one dimension and miss out on the benefits of the other. Paradoxical frames can also increase capacity to tolerate different perspectives and to integrate these perspectives into new linkages and creative ideas.

When we looked at the rigidity versus responsiveness through a paradoxical frame, we came up with the analogue of a top-down, controlling prison environment of a PMO versus a vibrant start-up-like studio or lab environment where people are engaged and creative within boundaries. We liked the discipline and consistency of the controlled environment, but we didn't want the rigidity. We liked the freedom and flexibility of the studio, but we didn't want a free-for-all. Instead of forcing people to follow PMO standards, we created a flexible environment for voluntary compliance, where they adopted PMO processes because they got value and enjoyed the experience. Viewing through a paradoxical frame helped us to create the idea of "rigor without rigidity," which was discussed in Chapter 2.

Another dilemma that project teams and PMOs have to deal with is the duality dilemma between the creatives and the logicians. According to Amy Webb in her book, *The Signals Are Talking,* she describes how you have probably experienced this clash in a meeting where the creative people felt as though their contributions were being discounted, while the logical thinkers—whose natural talents are in managing processes, projecting budgets, or mitigating risk—felt undervalued because they weren't coming up with bold new ideas. She recommends a technique taught at the Hasso Plattner Institute of Design at Stanford University (also known as "the d.school") that addresses the duality dilemma and illuminates how an organization can harness both strengths in equal measure, alternately broadening ("flaring") and narrowing ("focusing") its thinking. She explains:

> When a team is flaring, it is sourcing inspiration, making lists of ideas, mapping out new possibilities, getting feedback, and thinking big. When it is focusing, those ideas must be investigated, vetted, and decided upon. Flaring asks questions such as "What if?"; "Who could it be?"; "Why might this matter?"; and "What might be the implications of our actions?" Focusing asks "Which option is best?"; "What is our next action?"; and "How do we move forward?"

Finding the sweet spot between the paradoxes is a delicate balancing act. You have to learn to tip and tweak just the right amount. Start-ups that eschew rules altogether can get tangled in complexity, and the heavy structure and rules of incumbents makes them slow and stodgy. It is important to balance and counterbalance. If you are an established company weighing more on the traditional side, you might want to swing more to the next generation. Once you have reaped the benefits of flexibility, you can tilt to the other side as you strive to find your edge. Good leaders can sense which side to tilt and how much; they know how to distinguish between the complicated and complex and which approach to use, as they balance and counterbalance along the complexity continuum. They learn how to juxtapose the opposing forces of the need for rigor and flexibility and use it to create a dynamic loop for strategy execution.

LEARNING TO DANCE

Projects, organizations, and life are complex adaptive systems, and as we have discussed in this book, they are characteristic of the **DANCE**—the **D**ynamic and changing, **A**mbiguous and uncertain, **N**onlinear, **C**omplex, and **E**mergent and unpredictable in nature. To deal with the DANCE, you have to balance between SPEC and SRAA.

The traditional methods are based on scope, plan, execute, and control (SPEC). SPEC is good for linear, well-defined stable situations, and SPEC can help to manage the normal variance that is part of any project or program. But in some situations, no matter how hard you work to meticulously manage projects or PMOs, the more you try to control, the more it fluctuates and the harder it gets. You experience the instability caused by the DANCE, which swings the project beyond the thresholds of normal variance. This happens due to a combination of factors like the ambiguity of scope, the sheer number of linkages and dependencies, or the multiplicity of stakeholders involved. Scope, requirements, and solutions are emergent in nature and are hard to pin down and plan for in a continually shifting landscape. When you are dealing with one or more of these characteristics, you have to recognize that it is a different game. You need to learn to balance and manage the DANCE.

To manage the unexpected, an organic approach is required. You must cultivate skills to sense, respond, adapt, and adjust (SRAA). Sensing skills help to sense and see things that are hard to see and create unique insights; to develop acute awareness and vigilance to anticipate unexpected changes. Response prepares you to view the unique situation and respond accordingly at that moment. Adaptation helps you to quickly adapt to the new reality, and adjust the plan to accommodate the changing reality.

For example, you can plan, but don't get comfortable with it. Continue to ask penetrating questions and challenge the assumptions throughout the project lifecycle. Plans should be fluid and enable, not rigid and confine. Rigid plans in a DANCE environment can create blind spots that prevent you from seeing the unfolding project reality. Fluid plans enable you to sense and be open to emerging stakeholder needs and respond to unexpected changes.

SRAA cannot replace traditional scope, plan, execute, and control processes, in addition to good risk and change management practices. But when used in conjunction with those techniques, SRAA is instrumental in helping you manage the DANCE. When you are in the middle of your project, frustrated with unexpected fluctuations, you have to sometimes step back from the dance floor to see the big picture, recognize the DANCE, and manage its impact in your project environment. SRAA is recursive, and the order of sense, respond, adapt and adjust is going to be dynamic and not necessarily in that order. Think of SRAA as an operating system for agile. Without cultivating SRAA, it is hard to practice true agile, or go beyond agile to adaptive, emergent, and intuitive approaches.

Does DANCE Flow Better with Classical or Jazz Mindset?

Music sets the ambiance, mood, and context for dance. The question is what kind of music approach works better in a DANCE-world. The traditional management approach is akin to a classical orchestra with the project or program manager as the conductor. To deal with the DANCE you need the mindset of a jazz ensemble. You have to sense and perceive and improvise in the moment. You create order with little or no blueprint. There is minimal hierarchy and dispersed decision making, the opposite of a top-down classical approach. Jazz enables the practice of freedom with fences, where diverse specialists bring different perspectives, and they experiment, improvise, and innovate. It enables the emergence of soloing and fostering independent action while supporting and enabling the network.

AGILE BEING AGILE

The ideas in this book aimed to prepare your operating system (OS) for agile, so you can "be" agile before you "do" agile, as you apply the DNA of strategy execution.

While traditionalists and agilists debate on either end of the spectrum, cultivating a hybrid mindset seeks to achieve the balance between the two extremes.

Traditionalists believe in top-down heavy methodologies and processes and dismiss the value of agile practices. Agilists swear by the agile movement that advocates a responsive and iterative approach. There is even a budding community of post-agilists who have come full-circle after adapting agile practices. It must be emphasized that it is not that one is better than the other; it would be like comparing which is better, a car or a boat—each has its purpose.

As agile has become mainstream in recent years, beyond software development, there is a lot of discussion about the pros and cons, success stories, and implementation challenges.

A healthcare organization we worked with struggled to implement agile over the past year; in retrospect, they discovered that they were too caught up in figuring out which agile methodology to use as well as the associated tools and mechanics. Management and teams were confused as they were not prepared and did not understand the true spirit of agile. We come across many similar examples in agile environments where the project managers and teams feel good about using agile methods, but management and customers do not necessarily see the results. A classic case of focusing on the apps, without cultivating the right operating system.

In today's DANCE-world, the choice is not either/or, but a need to use the right mix and achieve a balance between the extremes. One way to strike a balance is to play to the strengths of each approach. A hybrid mindset goes beyond the mechanics of the approach toward facilitating and enabling results, outcomes, and changing customer needs and expectations.

As a project manager, you might be already adept at using agile practices in your organization. Others from traditional organizations might question how to apply agile, especially if that organization does not support it. Either way, cultivating a hybrid mindset can help you focus on the right things and become more successful.

Remember, there is no one best approach. It is up to you to find the "sweet spot" for yourself and your organization as you try to achieve balance between traditional and agile. As I have argued in this book, which approach to use depends on which game you are playing, pool or pinball, building a table, or planting a garden. Although both can benefit from agile, you just can't win at pinball with the traditional approach. In today's DANCE-world, chances are more and more you are doing the latter, which needs more of an agile mindset. This debate is gradually fading; in the age of agile, you have to be agile and know how to apply the right mix of methods, processes, and tools that are appropriate for your business, organization, culture, and nature of projects. The aim should be to move from process agility in execution to strategic agility and agility in each of the other areas of the DNA—governance, connection, measurement, change, and learning.

Beyond Agile: What's Next?

What's next after the predictive approaches of the mechanical and linear mindset, to the progressive elaboration and rolling wave approaches of iterative, incremental and agile approaches suitable for an organic world. As we discussed at the beginning of the book, today we are not dealing with waves anymore, it is more like being swept by disruptive tsunamis.

Recently, I got the opportunity to work with the PMO team of Intuitive Surgical, a pioneer in the rapidly emerging field of robotic-assisted minimally invasive surgery. It was quite an experience to get a first-hand experience what's possible with Da Vinci, the companies' leading surgical robot. Robotic-assisted surgery allows surgeons to perform many types of complex procedures with more precision, flexibility, and control than is possible with conventional techniques. Similarly, how can we develop and hone our skills that augment our capabilities with next-generation tools and technology? As technology gets more and more integrated with humans, we have to develop skills to work better with it. Matthew Milan writes about the centaur, a Greek myth about a half horse and half man, in a Fast Company, Co.Design article, *"More recently, the centaur has been used to describe human and artificial intelligence working together toward a common goal. This is the inevitable march of technology. Computers are faster and smarter than ever, and they will only improve, but they lack key cognitive skills like common sense and the ability to draw on a diverse set of experiences–things people do well. People and computers can be more effective working in tandem."*

In a world of ever-increasing technology with AI, bots, smart connected devices (SCDs), and augmented reality (AR), next-generation project/program managers, and PMOs have to learn to enhance their intuitive and integrative skills. By combining the strengths of humans and machines you can dramatically augment results and impact. This approach comes full circle to combine and connect the mechanical and organic paradigms that are necessary to thrive on the edge of chaos.

THRIVING AT THE EDGE OF CHAOS

The edge of chaos is an energizing and vibrant place between complexity and chaos where opportunity lurks. It is a state of dynamic balance between order and disorder. If you can get comfortable with the energizing discomfort at the edge, you can spot opportunities for creativity and innovation. You can see things that other might miss and prepares you to take smart risks. You can't relax for too long at the edge of chaos, but it makes you stronger.

Typically, project managers' and PMOs' job is to avoid risks or prevent failures, but this can also make the organization weak, as it is not prepared to withstand the turbulence caused by the DANCE. Best-selling author and professor Nassim Taleb, in his book *Antifragile: Things that Gain from Disorder*, writes:

> Systems that are antifragile: they don't just survive failures and external shocks, they get stronger as a result. Antifragility is beyond resilience or robustness. The resilient resists shocks and stays the same; the antifragile gets better.

The project team and PMO have to challenge itself and question whether they are creating a static and inert environment that cannot withstand the DANCE, or they are fostering flexibility and strength that can grow stronger with turbulence. Traditional project and PMO leaders seek order, control, and equilibrium, which can provide a false sense of security in a DANCE-world. They don't realize that their methods and approaches are making their environment and projects weaker. Instead, they should add stressors and variability so that the teams can learn and grow stronger from it.

On the other hand, inexperienced next-generation leaders can also run their organizations into burnout and chaos. They have to learn to craft and practice the barbell strategy, as explained by Nassim Taleb in *Antifragile:*

> The barbell (a bar with weights on both ends that weight lifters use) is meant to illustrate the idea of a combination of extremes kept separate, with avoidance of the middle. In our context it is not necessarily symmetric: it is just composed of two extremes, with nothing in the center. One can also call it, more technically, a bimodal strategy, as it has two distinct modes rather than a single, central one.
>
> I initially used the image of the barbell to describe a dual attitude of playing it safe in some areas (robust to negative Black Swans) and taking a lot of small risks in others (open to positive Black Swans), hence achieving antifragility. That is extreme risk aversion on one side and extreme risk loving on the other, rather than just the "medium" or the beastly "moderate" risk attitude that in fact is a sucker game (because medium risks can be subjected to huge measurement errors). But the barbell also results, because of its construction, in the reduction of downside risk—the elimination of the risk of ruin.

The balancing act is an art to thrive at the edge of chaos, where instead of drowning in the DANCE, you are comfortable surfing on the edge, sensing, responding, adjusting, and adapting. It's not easy to learn to DANCE—it takes a lot of patience and experience. With practice, though, you can realize that the norm for leaders in these situations is to get comfortable being uncomfortable with the DANCE!

WHAT IS THE NEXT GENERATION?

A question I get asked is, "You have been talking about the next generation for over a decade; when is it coming?" My response is that the next generation is never coming. One of the attendees from Nike corrected me and said that at Nike, "we believe it is always coming!" The point is that the next generation is not a destination, it is a journey without a finish line. Once you get from A to B, you need to move to C, and when you arrive at C, you should start thinking about D because if you don't, your competition is probably already there. It is a process in which you are never done; you execute, learn, balance, and tweak as the situation evolves and new information emerges. The next generation is a question: perpetual state of questioning and reinvention.

A good mantra for the next-generation project, program, and PMO managers is ABCD: Always Be Constructively Dissatisfied. Always questioning, inquiring, challenging … is there a better way … how can we do better?

The next-generation PMO is better prepared to deal with the DANCE and disruption because it pushes strategy execution to the edge, by constantly questioning and creating a culture of ABCD

The next-generation project manager and PMO have to be the bridge between the old and new. The traditional must be sustained to ensure stability while preparing the organization for change and agility. It is akin to rocking the boat while keeping it steady. The key is to constantly challenge yourself—how can we better adapt to changing needs, how can we better serve our customers and organization, and how can we add greater value and create long-term sustainable impact?

In this book, I have used the language of genes and DNA to decode strategy-execution, and PMOs. In complex adaptive systems, feedback loops are critical elements based on which genes survive and mutate. Another way to think about the next generation is with the lens of memes, which self-replicate based on communication and feedback loops. Successful, healthy, vibrant next-generation projects and PMOs self-replicate and mutate like memes based on positive experience. If you want to take your project and PMO to the next level, think about how you can constantly balance and strive for a positive experience, value, and greater impact.

TEN SKILLS FOR NEXT-GENERATION PROJECT MANAGEMENT AND PMOS

Following is a summary of the 10 skills necessary for next generation project management and PMOs based on the themes of this book. These are also the next generation of emergent and intuitive skills which might give us an edge over computers, at least for a short while. It is going to be rare for any one individual or team to excel in all of these skills. Use the list to assess your strengths and weaknesses, and see which skills you want to explore and develop. Who knows, you might even become a pinball wizard. More importantly these skills will help you to dance and thrive on the edge of chaos:

DANCE-ing. Recognize and understand the DANCE. Learn to dance and thrive on the edge of chaotic volatility, ambiguity, and unpredictability. Seek persistent disequilibrium and be comfortable being uncomfortable.

SRAA. Sense-respond-adapt-adjust mindset; sensing, and seeing and perceiving what others don't see. The presence and acute awareness to respond, adapt, and adjust to the changing reality with unique perspective and insight.

Strategic-Execution. Ability to link execution with strategy and connect performance with purpose. Execution of deliverables and outputs with results, outcomes, and winning mindset; a sense of purpose and making the right choices; ownership and commitment with a skin-in-the-game attitude and entrepreneurial spirit (Chapters 4 and 5).

Connecting. Identify the key connections and their interplay. Bridge and connect stakeholders, silos, business, and interfaces and interdependencies. Social skills to work through the hierarchy and expand their network. Identify leverage points and reinforce connections, and ensure the quality of communications that flow through these connections (Chapter 7).

Changemaking. Change intelligence to transform and make change happen. A changemaker isn't someone who simply manages change and wishes for change; he or she makes change happen. Changemakers have the ability to transform; they are not just game players or influencers, they are game changers. They influence the outcomes through responsibility, ownership, and determination (Chapter 9).

Learning and Seeking. Insatiable curiosity and learning agility; always questioning, inquiring, challenging … is there a better way? how can we do better? Learning from experimentation, trial and error, and experience. Developing applied curiosity and accelerated learning skills (Chapter 10).

Adaptive. Constantly seek and adapt to new information and changing circumstances. Rigor without rigidity, and openness to change and adaptation; finding the sweet spot between the extremes (Chapter 2).

Deep Generalist. Focus on getting broad expertise, becoming deep generalists, a term used by futurist Jamais Cascio, who explained in an interview:

> Learning a lot about a lot of things, and—just as important—getting a real understanding of how they are connected … what nature shows us is that the species that adapt best to radically changing environments are the generalists. But most generalists are shallow, living on the peripheries of more specialized ecosystems.

Deep generalists have their feet on the ground and their head in the clouds, they can zoom-in and out and thrive in many contexts.

Ownership. Think like an owner; extreme ownership of results and outcomes; bottom-up emergent leadership to direct and influence outcomes without title or authority to do so.

Artistry. Complement technical management skills with artistry; improvise based on experience, insight, intuition, and judgment and make appropriate adjustments. Think like a designer—seek to provide form, function, and structure to ensure the feasibility and viability to enable the vision. View constraints as creative opportunities for innovation Strive to be an artistic-mechanic (Chapter 5).

Whether it is the DANCE, SRAA, connecting, learning and seeking, or deep generalist skills, there is an underlying theme of developing capabilities for sensing and perceiving what others don't see. Project and program managers are typically good at the details and looking at things with a microscope. To develop next-generation skills, you have to not only be good at looking under the microscope but also know which lens to use and remind yourself to try different lenses that challenge existing mental models and provide a different perspective. Besides relying on microscopic views, next-generation leaders also use binoculars to zoom-out and see what is on the horizon. They also have telescopes in their toolkit to develop a long-term gaze and foresee future challenges and opportunities. You also need to embed periscopes as an extension of your eyes and ears for the things you cannot see from your level. Additionally, you also need to equip and augment yourself for an increasingly digital world with virtual reality (VR) and augmented reality (AR) tools and apps, to interact with the world in new ways and gain new perspectives. Next-generation project management and PMO leaders use and balance among all of these tools as they sharpen and practice the above skills.

APPLICATION QUESTIONS

One of my favorite parts of my talks and seminars is AMA: Ask Me Anything. Here are a few that you might be thinking about asking:

But wait, what's the big deal about the DANCE, couldn't we use risk management, a foundational knowledge area of project management to manage the DANCE?

Risk management is important and essential and should be practiced diligently with risk identification, assessment, and mitigation plans. But we also have to understand the limitations of traditional risk management. Risk management is effective to deal with known and known/unknown risks which are identifiable. It is ill-equipped to handle wicked risks – the unknown/unknowables, besides the unknown/unknowns. The first step of risk management is risk identification, if you can't even identify the risk, how can you manage it? Often in DANCE type of situations, traditional risk management can be dangerous, as it gives a false sense of security. Especially if you are unaware of the DANCE, and have confidence in your comprehensive risk register. Have you noticed that often the things that come back and bite us, are typically not on our risk list? It is not your fault, as a next-generation PM you should be aware that the 'E' in DANCE stands for emergent and unpredictable, and it was perhaps unknowable, and not possible to identify or predict. Instead of the classic risk identification, assessment, response, and control approach, you need a different approach to not only build resilience, and figure out how to make failure survivable, but go beyond and redesign for anti-fragility.

What is the validity of the DNA of strategy execution as a model?

The DNA framework emerged and has been fine-tuned over the past 15 years from working with a few thousand project and program managers, PMO leaders in the next-generation PMO and portfolio management, managing the DANCE, and leadership seminars. Ideas based on the DNA of strategy execution have been applied in many organizations in different industries around the world.

Working with PMOs, there continues to be confusion and disconnects regarding the purpose, role, and functions of a PMO. Also, the disconnect between strategy and execution, as well as the other areas of DNA, are a pervasive pain point that perpetuates siloes. We found the DNA to be a common-sense and holistic approach to address these questions and also outline the purpose of a successful PMO.

You have probably heard the quote, "All models are wrong, but some are useful." It's not perfect and can be tweaked more. Nor is it a magical tool that will solve all your problems. Viewed through a different lens, it might not make sense, and you might want to adjust the appropriate language that makes sense to you. Many organizations around the world in different sectors are using it as a framework to define and communicate their organizational project management and PMO capabilities, as well as assess their missing elements and strengths and weaknesses.

The DNA covers a lot of different areas; are project management and the PMO trying to take over the organization?

By design, the DNA is holistic and covers a lot of different areas. No, the idea is not that project management, or PMO is going to take over. In fact, you might want to be careful to share the DNA framework, without intimidating people, unless they understand the intent. The idea is to connect the different elements and optimize the whole. Also, the DNA is designed to instill a sense of ownership and connection at every level. What needs to be explained and understood is that even if one of the elements is missing or weak, it is not complete and does not function optimally. You may not be responsible or accountable for all the areas, as long as all the areas are being addressed. The role of the project and PMO leaders is to link the different areas and reduce disconnects. The DNA provides the right perspective to make sure you are taking a holistic approach and not missing any elements.

I have no authority or power in my organization, how is it possible to apply these ideas?

Of course, having formal authority helps but does not guarantee success. There are a lot of examples of project managers and PMOs with a lot of authority, but they are operational execution oriented with limited influence and impact. On the other hand, with the right combination of competence, drive, attitude, and social competence, you can accomplish a lot more and earn a seat at the strategy table and drive strategic-execution. As the famous Gandhi quote says, "You have to be the change you want to see in the world." Start, by thinking like an owner and identifying the opportunities to influence and make an impact in your won sphere.

How do I start to apply the DNA of strategy execution?

The first thing is not to get overwhelmed by all the elements. Like a good doctor, start by diagnosing the pain and challenges. Pick one area and focus on it, and see what needs to be done to address it. As you focus on the one element, whether it is execution or governance or measurement, or any of the other DNA areas, the key is also to think holistically and connect to the rest of the six elements and what needs to be done to optimize the whole. You can also assess and review the strengths and weaknesses and see where you can add value as you chart your road map. As you are implementing, remember the importance of measurement, feedback, change, and communication along the way.

You don't understand my organization and culture—these next-generation ideas are never going to work in my company.

You are right; I am not sure if these ideas will work for your organization or not. It depends on your business, culture, politics, leadership, and many other factors. But that does not mean you cannot try and experiment in your own way, in your domain. The proof is in the pudding; if it works, people will notice the results and impact. The key is to keep questioning, collaborating, and experimenting as you test and observe your hypothesis of what works. Also, it pays to practice intelligent disobedience from time to time, to take risks and try next-generation approaches and surprise stakeholders with positive experience and impact.

> "I would rather have questions that can't be answered, than answers which can't be questioned."
>
> *Attributed to Richard Feynman, source unconfirmed*

FEEDBACK LOOPS: PROJECT MANAGEMENT/PMO DELIGHT INDEX (PDI)

To thrive in a DANCE-world, you have to adapt and evolve. Successful adaptation has three characteristics drawn from evolutionary biology according to Ronald Heifetz et al. in their book, *The Practice of Adaptive Leadership:* (1) it preserves the DNA essential for the species' continued survival; (2) it discards (reregulates or rearranges) the DNA that no longer serves the species' current needs; and (3) it creates DNA arrangements that give the species' the ability to flourish in new ways and in more challenging environments. For successful adaptation, you have to rely on feedback loops to learn and evolve.

Feedback loops are crucial to learn what needs adjustment. Projects, programs, and PMOs have to design feedback loops to ensure they are getting the right information from diverse channels to evolve and adapt. Part of the marcom design has to include a combination of surveys, informal feedback from key stakeholders, measures based on different project sensors, metrics, and a variety of social media channels.

Imagine how you could design your project and PMO experience to learn how to surprise and delight your customers and stakeholders by turning their pain points into delight points. As you simplify and make structure and processes easier from your endusers' and customers' perspective. You will surprise your stakeholders with the positive experience of dealing with the PMO. They will be compelled to share this experience with others and make the PMO viral and self-replicating for various aspects across the organization.

It is important to measure and seek frequent feedback. Project managers and PMOs can gather valuable feedback from its customers and stakeholders by using a simple PM/PMO delight index (PDI) that covers the range of business management evolution, from efficiency and effectiveness to experience and impact. Adjust the language in the project/program and PMO delight index examples in Figures 12.1 and 12.3 to suit your purpose.

1. Efficiency: Did the project/program meet the defined delivery outputs and execution success criteria?	☹ 1 2 3 4 ☺
2. Effectiveness: Is the project/program achieving, or how likely it is to achieve defined objectives and key results or promised outcomes and desired benefits?	☹ 1 2 3 4 ☺
3. Experience:	
A. How would you rate the overall experience with this project/program?	☹ 1 2 3 4 ☺
B. How likely are you to share about this project/program?	☹ 1 2 3 4 ☺
4. Impact: How would you rate the overall impact of this project/program?	☹ 1 2 3 4 ☺
© J. Duggal. Projectize Group 2017	

Figure 12.2: Project/Program Delight Index
Source: © J. Duggal. Projectize Group.

1. Efficiency: Are the PMO processes/services/tools helpful in executing projects faster, better, cheaper?	☹ 1 2 3 4 ☺
2. Effectiveness: Did you achieve the desired project/ program results & outcomes using the PMO processes/ services/tools?	☹ 1 2 3 4 ☺
3. Experience:	
A. Was it easy and frictionless to work with the PMO?	☹ 1 2 3 4 ☺
B. How likely are you to share your PMO experience or get more involved with PMO activities?	☹ 1 2 3 4 ☺
4. Impact: How would you rate the overall value and impact of the PMO?	☹ 1 2 3 4 ☺
© J. Duggal. Projectize Group 2017	

Figure 12.3: PMO Delight Index
Source: © J. Duggal. Projectize Group.

BEN OR BOB?

Remember Ben and BoB from Chapter 8? While Ben is focused on delivery and outputs, BoB is looking at the big picture aiming for benefits and outcomes. Ben is satisfied with the near-term, what's in view, within reach, easily measurable benefit, which is often limited. Ben is worried about the form, BoB is after the essence. While Ben may be happy looking and analyzing the finger pointing to the moon, BoB is seeking the experience and glory of the moon.

The question is, who do you want be, Ben or BoB?

A quick response might be, "more like BoB," but as you reflect on the nuance, you realize you can't get to BoB without Ben, who is execution oriented, while BoB is strategy focused. Neither is better than the other; we need both. If you are more Ben oriented, you need to start to zoom-out and challenge yourself to see the world with BoB's eyes. If you are BoB, you have to understand you cannot realize your strategy without Ben, and develop some Ben capabilities, or make sure you have some Ben's on your team. The ideal is BobbyBen, who is bimodal, constantly bobbing and balancing between the complicated and the complex, toward the sweet spot at the edge of chaos. With the DNA of strategy execution, he/she can leverage all aspects of the DNA to achieve desired results and make an impact with strategic-execution. The choice is yours; you want to deliver forgettable projects and programs, or leave a legacy of memorable projects and programs whose experience and impact is felt long after they are over.

"The only true wisdom is knowing you know nothing."

Socrates

APPLICATION OF DNA OF STRATEGY EXECUTION

Use the following table as a checklist to assess, apply and develop the elements of the DNA of Strategy Execution. This is oriented for projects and programs, but can be adapted as a general framework for management and effective strategy-execution. (Organized by DNA Elements and Strands)

Strategy	Execution	Governance	Connect
Strategy and Business Model Alignment □**Diagnosis:** Assess and analyze the problem this project/program is addressing; business case alignment; Customer and stakeholder pain points and needs assessment □ **Choice:** Prioritize activities based on strategic choices; Choose and prioritize requirements □**Design:** Business case alignment rules based on strategy; connect and link initiatives, programs, and projects to test	**People** □Assess and develop team members with balanced next generation skills and capabilities (see Execution and Balance chapters for more on next generation skills) **Process** □Utilize relevant PM Processes— initiation, planning, execution, monitoring, and control and closeout **Knowledge Areas:** □Integration □Scope □Time □Cost □Quality □Human resources □Communications	**Steering** □Identify or establish appropriate steering structure for project/program **Standards** □Utilize established standards and methodologies **Policies–Procedures** □Identify, develop, or adopt appropriate polices, and procedures for each of the DNA areas of strategy, execution, connect, measure, change, and learn **Gates** □Review and prepare for stage-gate criteria Participate in stage-gate reviews	**Customer and Stakeholder Management** □Stakeholder identification, classification, prioritization, and management □Identify and define customer □Facilitate stakeholder and customer engagement **Silos** □Identify and map relevant organizational silos related to project/program □Identify and map organizational matrix roles related to project/program

(continued)

Strategy	Execution	Governance	Connect
coherence and identify redundancies	□ Risk □ Procurement □ Stakeholder mgt.	**Review–Audit** □ Prepare for project/ program reviews Participate in project / program reviews and audits	**Business** □ Review and identify business model activities □ Identify and connect project/program activities to business model □ Review or develop project business model canvas
□ **Action:** Review and validate business model canvas; business case review and alignment; provide strategic decision-support	**Technology: Tools, Systems, Apps, and Bots** □ Utilize and leverage appropriate combination of tools and systems	**Compliance** □ Review and complete compliance requirements for: Regulatory	
□ **Evolve:** Retrospective, learning, iteration, and evolvement	**Flow** □ Map critical workflows	□ Legal □ Financial □ Security	**Interfaces and Interdependencies** □ Identify and map org. interfaces
Strategic Risk and Investment Management	□ Identify interfaces □ Identify and map interdependencies □ Identify bottlenecks □ Optimize workflow	□ Environmental □ Industry specific □ International	□ Identify and map cross-project interdependencies □ Connect interfaces and interdependencies
□ **Diagnosis:** Risk and assumption analysis		**Responsibility and Accountability**	
□ **Choice:** Boundaries for risk acceptance, avoidance, sharing		□ Establish/clarify responsibility and accountability (RACI)	**Network and Connections** □ Map stakeholder formal and informal networks
□ **Design:** Risk identification, assessment, risk response, and control approach		**Authority** □ Identify, review, clarify relevant authority thresholds	□ Stakeholder network analysis □ Identify structural holes, gaps, and corrective actions
□ **Action:** Proactive risk review and management; environmental scanning		**Decision-rights** □ Identify, review, and clarify relevant decision-rights	**Marcom** □ Develop marcom and branding strategies for project/program Promote project/ program visibility, benefits, and value
□ **Evolve:** Retrospective, learning, iteration, and evolvement		**Rules–Guidelines** □ Identify, review, clarify, and communicate— boundary, prioritizing, and stopping rules or guidelines	□ Monitor and seek communications feedback □ Identify communications gaps
Benefits, Value, and Impact Management			□ Facilitate corrective actions
□ **Diagnosis:** Assess and evaluate, results, benefits, value, and impact analysis			

Strategy	Execution	Governance	Connect
□ **Choice:** Determine what approaches and actions will move the needle and identify opportunities for benefits optimization □ **Design:** Benefits management approach—identification, mapping, alignment, prioritization, realization, and benefits realization milestones (BRMs) □ **Action:** Benefits management—Identify opportunities for benefits optimization; Review and recommend actions for cost savings, cost avoidance, and optimization □ **Evolve:** Results, benefits, value, and impact assessment and analysis; Retrospective, learning, iteration, and evolvement			**Relationship Management** □ Identify and develop relationships with key stakeholders □ Identify and resolve bottlenecks and organizational interfaces □ Business partner relationship management (partner with customers to understand business/project needs) □ Vendor and outsourcing management **Community** □ Participate and facilitate collaboration in relevant communities of practice (internal and external as appropriate)

Measure	Change	Learn
Objectives	**Awareness**	**Culture**
□ Define and clarify success □ Define and clarify measure purpose and perspective □ Define and clarify objectives □ Define and establish key results □ Establish customer-centric and stakeholder-centric measures	□ Cultivate importance of change □ Promote awareness of change adoption lifecycle □ Identify and highlight change risks □ Develop configuration mgt. plan □ Facilitate change advisory board (CAB)	□ Cultivate project/program learning culture □ Practice execution as learning □ Make learning fun and rewarding

(continued)

Measure	Change	Learn
Key Results (Measures/Metrics) ☐ Define outcome (BoB) measures ☐ Define output (Ben) measures ☐ Identify what behaviors lead to outputs ☐ Identify what behaviors lead to outcomes ☐ Balance output and outcome measures ☐ Create and maintain metrics profile ☐ Ensure metrics alignment to business objectives **Reporting** ☐ Design and develop information radiators, reports, dashboards, and scorecards ☐ Develop, review, maintain, and track status, health, progress, and trend reports. ☐ Develop, review, maintain, and track executive dashboards ☐ Gather, consolidate, publish, and distribute reports **Action** ☐ Review and analyze dashboards and reports ☐ Assess progress in achieving business objectives through the projects underway ☐ Identify and respond to weak or troubled project/program performance ☐ Conduct project/program phase reviews ☐ Implement post-project (closeout) reviews, post-mortems, or retrospectives ☐ Plan and facilitate corrective actions	**Anticipation** ☐ Assess change anticipation capabilities ☐ Assess DANCE factors ☐ Scan PESTLE factors ☐ Assess behavioral outcomes **Absorption** ☐ Assess customer and stakeholder absorption capacity: ☐ Volume of change ☐ Velocity of change ☐ Complexity and impact of change ☐ DICE factors ☐ Assess and develop change readiness plan **Adoption Customer** ☐ Assess and plan behavioral adoption ☐ Customer impact assessment (CIA) ☐ Customer and stakeholder empathy mapping ☐ Implement and validate customer adoption metrics **Choice** ☐ Assess, review, and design choice architecture and nudge factors ☐ Utilize structured checklists **Communication** ☐ Effective marcom: craft engaging, inspiring, and resonating project/program change messaging ☐ Communicate change expectations and success criteria **Connectors** ☐ Engage different levels of stakeholders ☐ Conduct stakeholder network analysis	**Curiosity** ☐ Cultivate and promote curiosity and learning ☐ Explore ways to shift from work-place to learn-place ☐ Inspire and promote questioning and reflection **Capture** ☐ Explore and implement feedback loops, feed-forward, retrospectives, pre-mortems, storytelling ☐ Utilize established KM tools to capture knowledge artifacts, and disseminate and share lessons learned and best practices **Community and Collaboration** ☐ Participate in learning communities of practice ☐ Promote and spark collaboration ☐ Utilize established collaboration platforms and tools **Curation** ☐ Identify, organize, and disseminate—project/program artifacts, ideas, lessons learned, and best practices **Correcting** ☐ Document and capture failure, mistakes and issues ☐ Analyze failure and mistakes and develop antidotes to avoid repeat failures **Continuous Innovation** ☐ Practice continuous improvement ☐ Apply double-loop learning to evolve from continuous improvement to continuous innovation opportunities

Measure	Change	Learn
Learning (Feedback) ☐ Review and analyze metrics effectiveness ☐ Seek reporting feedback ☐ Design and implement feedback loops ☐ Update and fine-tune reporting effectiveness	☐ Identify, support, and amplify positive deviants, influencers, and opinion leaders for promoting project/program objectives, deliverables, and outcomes	

B
PMO FUNCTIONS AND ACTIVITIES SERVICE CATALOG

Use this service catalog as a checklist to identify, assess, and develop PMO Functions and Activities (Organized by DNA Elements and Strands)

Strategy	Execution	Governance	Connect
Strategy and Business Model Alignment	**People**	**Steering**	**Stakeholder Management and Customer Identification and Engagement**
□ **Diagnosis:** Assess and analyze project/program business case alignment; customer and stakeholder pain points and needs assessment	□ Talent management and professional development	□ Facilitate definition and establishment of steering board or committee	□ Stakeholder identification, classification, prioritization and management
□ **Choice:** Facilitation and/or clarification of business strategy, goals, and objectives and related choices	□ Talent assessment and skills inventory	**Standards**	□ Identify and define customer
□ **Design:** Craft business case alignment rules based on strategy; connect and link initiatives, programs, and projects to test coherence and identify	□ Design, develop PM learning and development approach	□ Review, select and adopt PM standards:	□ Facilitate stakeholder and customer engagement
	□ Provide training and professional development	□ Project/Program/Portfolio management	
	□ Develop and facilitate PM certification and career path approach	□ Review, select, and adopt industry-specific standards and methodology	**Silos**
	□ Coaching and mentoring	□ Establish internal standards for requirements, estimation, reporting, utilization, escalation, etc.	□ Identify and map organizational silos
	□ PM coaching and development		□ Identify and map organizational matrix roles
	□ Mentoring program		□ Define and clarify matrix roles

(continued)

Strategy	Execution	Governance	Connect
redundancies and design appropriate boundaries and criteria □ **Action:** Review and validate business model canvas; business case review and alignment; provide strategic decision-support □ **Evolve:** Retrospective, learning, iteration, and evolvement **Portfolio Management and Prioritization Diagnosis:** Portfolio review and analysis □ **Choice:** Facilitation and clarification of portfolio governance and boundaries □ **Design:** Classification, selection, prioritization and balancing—criteria and models □ **Action:** Business case review, ranking, selection, prioritization, balancing, termination □ **Evolve:** Retrospective, learning, iteration, and evolvement **Strategic Risk and Investment Management Diagnosis:** Risk and assumption analysis □ **Choice:** Boundaries for risk acceptance, avoidance, or sharing	□ Upskill experienced PMs through coaching **Process** □ Review, assess, dev, and support: **PM Processes—** initiation, planning, execution, monitoring, and control and closeout □ Knowledge areas: □ Integration □ Scope □ Time □ Cost □ Quality □ Human resources □ Communications □ Risk □ Procurement □ Stakeholder mgt. **PM Support** □ Project start-up workshops □ Planning workshops □ Risk management workshops □ Organizational readiness review □ Project review and guidance □ Lessons learned and retrospective facilitation □ Troubled project recovery **Technology: Tools, Systems, Apps and Bots** □ Identify, select, review and implement PPM and other tools and apps □ PPM admin and support	**Policies– Procedures** □ Review, assess, establish, fine-tune, and standardize processes for critical: □ Execution activities and Relevant strategy, connect, measure, change and learn **Gates** □ Develop / facilitate stage gate review process and criteria □ Conduct / facilitate stage gate reviews **Review – Audit** □ Conduct / facilitate project / program reviews □ Assist prep for project / program audits □ Facilitate project / program audits **Compliance** □ Review, assist, and support: compliance for: □ Regulatory □ Legal □ Financial □ Security □ Environmental □ Industry specific □ International **Responsibility and Accountability** □ Establish/clarify responsibility and accountability (RACI)	□ Connect and bridge silos and org matrix roles **Business** □ Identify business model activities □ Identify organizational activities □ Connect business activities and org. priorities **Interfaces and Interdependencies** □ Identify and map org. interfaces □ Identify and map cross-portfolio interdependencies □ Connect interfaces and interdependencies **Network and Connections** □ Map stakeholder formal and informal networks □ Stakeholder network analysis □ Identify structural holes, gaps, and corrective actions **MarCom** □ Marketing, communications, and branding PMO services □ Marketing, communications, and branding support for key initiatives, projects, and programs □ Promote PM awareness throughout organization

Strategy	Execution	Governance	Connect
□ **Design:** Risk identification, assessment, risk response, and control approach	□ Tools training and support	**Authority**	□ Monitor and seek communications feedback
□ **Action:** Proactive risk review and management; environmental scanning	**Flow**	□ Clarify/define and communicate PM and PMO authority:	□ Identify communications gaps
□ **Evolve:** Retrospective, learning, iteration, and evolvement	□ Map critical workflows	□ Business	□ Facilitate corrective actions
	□ Identify interfaces	□ Functional	
Resource and Capacity Management Diagnosis: Resource capacity and demand management analysis	□ Identify and map interdependencies	□ Financial	**Relationship Management**
	□ Identify bottlenecks	□ Contractual	□ Identify and develop relationships with key stakeholders
	□ Optimize workflow	□ Oversight	□ Identify and resolve bottlenecks and organizational interfaces
□ **Choice:** Resource procurement and management strategy		**Decision-rights**	□ Business partner relationship management (partner with customers to understand business/project needs)
□ **Design:** Effective resource allocation and capacity planning; dependency analysis across portfolio		□ Establish/clarify and communicate decision-rights	
		Rules–Guidelines	□ Vendor and outsourcing management
□ **Action:** Resource estimation, allocation and leveling; identify resource overloads; resource capacity commitment; assess organizational capacity		□ Establish/clarify and communicate— boundary, prioritizing, and stopping rules or guidelines	□ Manage customer demand for PMO services
			□ Filling and managing organizational gaps across business units and silos
□ **Evolve:** Retrospective, learning, iteration, and evolvement			**Community**
			□ Facilitate PM communities of practice
			□ Organize and support community

(continued)

Strategy	Execution	Governance	Connect
Benefits, Value and Impact Management Diagnosis: Results, benefits, value, and impact analysis			
□**Choice:** Determine what approaches and actions will move the needle and identify opportunities for benefits optimization □**Design:** Benefits management— identification, mapping, alignment, prioritization, realization □**Action:** Benefits management— review and recommend actions for cost savings, cost avoidance, and optimization □**Evolve:** Results, benefits, value, and impact assessment; retrospective, learning, iteration, and evolvement			

Measure	Change	Learn
Objectives	**Awareness**	**Culture**
□Define and clarify success □Define and clarify objectives □Define and establish key results □Define and clarify measure perspective □Establish customer-centric and stakeholder-centric measures	□Cultivate importance of change □Promote awareness of change adoption lifecycle □Identify and highlight change risks □Facilitate configuration mgt. □Facilitate change advisory board (CAB)	□Cultivate learning culture □Promote execution as learning □Make learning fun and rewarding

Measure	Change	Learn
Key Results (Measures/Metrics) ☐ Define outcome (BoB) measures ☐ Define output (Ben) measures ☐ Identify what behaviors lead to outputs ☐ Identify what behaviors lead to outcomes ☐ Balance output and outcome measures ☐ Create and maintain metrics profile ☐ Ensure metrics alignment to business objectives **Reporting** ☐ Design and develop information radiators, reports, dashboards, and scorecards ☐ Develop, review, maintain, and track status, health, progress, and trend reports ☐ Develop, review, maintain, and track executive dashboards ☐ Gather, consolidate, publish, and distribute reports **Action** ☐ Review and analyze dashboards and reports ☐ Assess progress in achieving business objectives through the projects underway ☐ Identify and respond to weak or troubled project/program performance ☐ Conduct project/program phase reviews ☐ Implement post-project (closeout) reviews; post-mortems or retrospectives ☐ Plan and facilitate corrective actions ☐ Monitor the performance of the PMO	**Anticipation** ☐ Assess change anticipation capabilities Assess DANCE factors ☐ Scan PESTLE factors ☐ Assess behavioral outcomes **Absorption** ☐ Assess customer and stakeholder absorption capacity: ☐ Volume of change ☐ Velocity of change ☐ Complexity and impact of change ☐ DICE factors ☐ Review portfolio pipeline and resource capacity ☐ Assess and develop change readiness **Adoption** **Customer** ☐ Assess and plan behavioral adoption ☐ Customer impact assessment (CIA) ☐ Customer and stakeholder empathy mapping ☐ Implement and validate customer adoption metrics **Choice** ☐ Assess choice architecture and nudge factors ☐ Develop structured checklists **Communication** ☐ Change education ☐ Effective marcom: Craft engaging, inspiring, and resonating change messaging ☐ Communicate change expectations and success criteria	**Curiosity** ☐ Cultivate and promote curiosity and learning ☐ Explore ways to shift from work-place to learn-place ☐ Inspire and promote questioning and reflection **Capture** ☐ Explore and implement feedback loops, feed-forward, retrospectives, pre-mortems, storytelling ☐ Identify, review, select, deploy, and maintain KM tools ☐ Collect and maintain knowledge artifacts ☐ Collect and disseminate lessons learned and best practices **Community and Collaboration** ☐ Cultivate learning communities of practice ☐ Promote and spark collaboration ☐ Explore and implement collaboration platforms and tools **Curation** ☐ Identify, organize, and disseminate—artifacts, ideas, lessons learned, best practices, tools, and apps ☐ Review and design knowledge mgt. taxonomy ☐ Curate new approaches to strategy-execution, innovation, knowledge sharing, and collaboration **Correcting** ☐ Document and capture failure, mistakes, and issues ☐ Analyze failure and mistakes and develop antidotes to avoid repeat failures ☐ Develop and improve checklists

(continued)

Measure	Change	Learn
Learning (Feedback)	**Connector**	**Continuous Innovation**
☐ Review and analyze metrics effectiveness ☐ Seek reporting feedback ☐ Design and implement feedback loops ☐ Update and fine-tune reporting effectiveness	☐ Engage different levels of stakeholders ☐ Conduct stakeholder network analysis ☐ Identify, support, and amplify positive deviants, influencers, and opinion leaders	☐ Review and promote continuous improvement ☐ Explore ways to evolve from continuous improvement to continuous innovation

BIBLIOGRAPHY

Chapter 1: Introduction: Strategy Execution in a DANCE-world

Benoit, Andy, "The Case for the Broncos," January 13, 2014. https://www.si.com/vault/2014/01/13/106417354/the-case-for-the-broncos.

Berger, Warren, *A More Beautiful Question: The Power of Inquiry to Spark Breakthrough Ideas*, Reprint edition (Bloomsbury USA, September 13, 2016).

Bevelin, Peter, *Seeking Wisdom: From Darwin to Munger*, 3rd edition (PCA Publications L.L.C., 2007).

De Toni, Alberto F., and Luca Comello, *Viaggio Nella Complessita* (Marsilio 2012).

Denning, Stephen, "Can Big Organizations Be Agile?" Accessed July 25, 2017. https://www.forbes.com/sites/stevedenning/2016/11/26/can-big-organizations-be-agile/#5ec2e16f38e7.

Denning, Stephen, "The Copernican Revolution in Management," Accessed July 25, 2017. https://www.forbes.com/sites/stevedenning/2013/07/11/the-copernician-revolution-in-management/#5d71024a108d.

Diamandis, Peter H., and Steven Kotler, *Bold: How to Go Big, Create Wealth and Impact the World*, Reprint edition (Simon & Schuster, February 23, 2016).

Drucker, Peter, Foreword to *Management,* Revised edition, by Jim Collins (Harper Business; April 22, 2008).

Drucker, Peter, *The Practice of Management* (Harper & Row, January 1, 1954).

Duggal, Jack, "The DANCEing PMO: Next Generation PMO for a Disruptive World." Paper presented at the PMI Global Congress North America Proceedings, Orlando, Florida, October 2015.

"Google's Larry Page on Why Moon Shots Matter," https://www.wired.com/2013/01/ff-qa-larry-page/.

"Manifesto for Agile Software Development," http://agilemanifesto.org.

Heifetz, Ronald A., Linsky, Marty and Grashow, Alexander, *The Practice of Adaptive Leadership: Tools and Tactics for Changing Your Organization and the World* (Harvard Business Press, May 18, 2009).

Miller, John H., and Scott E. Page, *Complex Adaptive Systems: An Introduction to Computational Models of Social Life* (Princeton University Press, 2007).

Peaucelle, Jean-Louis, and Cameron Guthrie, *Henri Fayol, the Manager* (Routledge, July 22, 2015).

Project Management Institute, *Agile Practice* (Newtown Square, PA: Project Management Institute, 2017).

"A Prescient Warning to Boeing on 787 Trouble." Accessed July 25, 2017. http://old
.seattletimes.com/html/sundaybuzz/2014125414_sundaybuzz06.

Ries, Eric, *The Lean Startup: How Today's Entrepreneurs Use Continuous Innovation
to Create Radically Successful Businesses,* 1st edition (Crown Business, September 13, 2011).

Rose, David S., *Angel Investing: The Gust Guide to Making Money and Having Fun
Investing in Startups,* 1st edition (Wiley, April 28, 2014).

Senge, Peter, *The Fifth Discipline: The Art and Practice of the Learning Organization,*
1st edition (Doubleday Business, August 1, 1990).

Stiehm, Judith Hicks, and Nicholas W. Townsend. The U.S. Army War College: Military
Education in a Democracy (Temple University Press, 2002).

Taleb, Nassim Nicholas, *Antifragile: Things That Gain from Disorder,* Reprint edition
(Random House Trade Paperbacks, January 28, 2014).

Wheatley, Margaret J., *Leadership and the New Science: Learning about Organization
from an Orderly Universe* (Berrett-Kohler, 1992).

Chapter 2: Agility: Rigor without Rigidity

Adams, Marilee G., *Change Your Questions, Change Your Life: 10 Powerful Tools for Life
and Work,* 3rd edition (Berrett-Koehler, January 11, 2016).

Berger, Warren, *A More Beautiful Question: The Power of Inquiry to Spark Breakthrough
Ideas,* Reprint edition (Bloomsbury USA, September 13, 2016).

Bevelin, Peter, *Seeking Wisdom: From Darwin to Munger,* 3rd edition (PCA Publications
L.L.C., 2007).

"Dirt Paths on Drillfield to Be Paved." Accessed July 25, 2017. http://www.vtnews.vt
.edu/articles/2014/08/080514-vpa-drillfieldpaths.html.

"The Global Innovation 1000: The Top Innovators and Spenders." Accessed July 25,
2017. https://www.strategyand.pwc.com/innovation1000.

Gregersen, Hal, "Use Catalytic Questioning to Solve Significant Problems," *Harvard Business
Review,* July, 19, 2013. https://hbr.org/2013/07/catalytic-questioning-five-ste.

"Harley-Davidson, Inc.—Optimizing Talent: A Culture of Empowerment." Accessed
July 25, 2017. http://www.catalyst.org/knowledge/harley-davidson-inc%E2%80
%94optimizing-talent-culture-empowerment.

Hobbs, Brian, and Monique Aubry, *The Project Management Office (PMO): A Quest for
Understanding* (Project Management Institute, 2010).

Kim, W. Chan, and Renée Mauborgne. *Blue Ocean Strategy: How to Create Uncontested
Market Space and Make the Competition Irrelevant,* Expanded edition (Harvard Business
Review Press, January 20, 2015).

Klein, Gary, *Seeing What Others Don't: The Remarkable Ways We Gain Insight* (Public
Affairs, June 25, 2013).

"Least Resistance: How Desire Paths Can Lead to Better Design." Accessed July 25, 2017. http://99percentinvisible.org/article/least-resistance-desire-paths-can-lead-better-design/.

Martin, Roger, *The Opposable Mind: How Successful Leaders Win Through Integrative Thinking* (Harvard Business School Press, 2007).

Martin, Roger, and A.G. Lafley, *Playing to Win* (Harvard Business Review Press, 2013).

Morgan, Adam, and Mark Barden, *A Beautiful Constraint: How to Transform Your Limitations into Advantages, and Why It's Everyone's Business* (Wiley, January 20, 2015).

PMI's Pulse of the Profession, 9th Global Project Management Survey, 2017.

Questions for Charles O'Reilly, "Organizational Ambidexterity," *Rotman Magazine,* University of Toronto, Fall 2014.

Thibaut, John, and Laurens Walker, *Procedural Justice: A Psychological Analysis* (Lawrence Erlbaum Associates, 1975).

"UC Berkeley Purposely Waited to Put in Paths after Seeing Where Foot Traffic Created Them." Accessed July 25, 2017. https://www.reddit.com/r/DesirePath/comments/23a6mi/uc_berkeley_purposely_waited_to_put_in_paths/.

Watson, Richard, "Six Sigma and Innovation Culture," *Fast Company,* December 2, 2007. https://www.fastcompany.com/661292/six-sigma-and-innovation-culture.

Wenger, Etienne, *Communities of Practice: Learning, Meaning, and Identity,* 1st edition (Learning in Doing: Social, Cognitive and Computational Perspectives) (Cambridge University Press, September 28, 1999).

Chapter 3: DNA of Strategy Execution

"A Danny Meyer Dictionary." Accessed July 31, 2017. http://query.nytimes.com/gst/fullpage.html?res=9E05E2DB1E31F936A15757C0A9669D8B63.

Grenny, Joseph, Kerry Patterson, David Maxfield, Ron McMillan, and Al Switzler, *Influencer: The New Science of Leading Change,* 2nd edition (McGraw-Hill Education, May 14, 2013).

Meyer, Danny, *Setting the Table: The Transforming Power of Hospitality in Business* (Harper Perennial; Reprint edition, January 29, 2008).

"Performance with Purpose." http://www.pepsico.com/Purpose/Performance-with-Purpose.

Project Management Institute, *A Guide to the Project Management Body of Knowledge* (PMBOK® guide), 6th edition (Newtown Square, PA: Project Management Institute, 2017).

Project Management Institute, *Organizational Project Management Maturity Model (OPM3),* 3rd edition (Newtown Square, PA: Project Management Institute, 2013).

Project Management Institute, *PMI's Pulse of the Profession®: PMO frameworks* (Newtown Square, PA: Project Management Institute, 2013).

Project Management Institute, *The Standard for Program Management,* 4th edition (Newtown Square, PA: Project Management Institute, 2017).

Projectize Group, LLC, USA, Project/Program Management Office (PMO) Survey (2005–2017).

Quinn, James, "CEO Spotlight: Jeff Bezos, One Eye on the Consumer, the Other on the Future," *Rotman Magazine,* University of Toronto, Spring 2016.

Schoen, Mbula, *Four Types of PMOs That Deliver Value,* June 17, 2016. https://www .gartner.com/doc/3345135/types-pmos-deliver-value.

Chapter 4: Strategy

Association for Project Management, *APM Winning Hearts Brochure* (Benefits Management Special Interest Group, 2011).

"Audi's WEC Track Record at Le Mans (since 2012)." Accessed July 25, 2017. https:// www.audi.co.uk/about-audi/events-and-partnerships/le-mans.html.

Boland, Richard Jr., and Fred Collopy, "Design Matters" in *Managing as Design* (Stanford Business Books, July 28, 2004).

Boston Consulting Group (BCG), "The Product Portfolio." https://www.bcgperspectives .com/content/classics/strategy_the_product_portfolio/.

Filippov, Sergey, et al., *The Strategic Role of Project Portfolio Management: Evidence from the Netherlands* (Proceedings of the 7th International Conference on Innovation & Management, 2010).

Ito, Joe, and Jeff Howe, *Whiplash: How to Survive Our Faster Future* (Grand Central Publishing, 2016).

Keeley, Larry, and Ryan Pikkel, *Brian Quinn and Helen Waters, Ten Types of Innovation—The Discipline of Building Breakthroughs* (Wiley, 2013).

Lafley, A.G., and Roger L. Martin, *Playing to Win: How Strategy Really Works* (Harvard Business Review Press, February 5, 2013).

Leinwald, Paul, and Cesare Mainardi, "Creating a Strategy that Works," *Strategy+Business,* Spring 2016.

Markowitz, Harry, "Portfolio Selection," *Journal of Finance,* 1952.

Martin, Roger, "Don't Let Strategy Become Planning," *Harvard Business Review, February 5, 2013.*

Moore, Geoffrey, *Zone to Win: Organizing to Compete in an Age of Disruption* (Diversion Publishing, November 3, 2015).

Morgan, Adam, and Mark Barden, *A Beautiful Constraint: How to Transform Your Limitations into Advantages, and Why It's Everyone's Business* (Wiley, January 20, 2015).

Osterwalder, Alexander, and Yves Pigneur, *Business Model Generation: A Handbook for Visionaries, Game Changers and Challengers* (Wiley, July 13, 2010).

Pascale, Richard, Mark Milleman, and Linda Gioja, *Surfing the Edge of Chaos: The Laws of Nature and the New Laws of Business* (Crown Business, October 17, 2000).

Porter, Michael, "What Is Strategy," *Harvard Business Review,* November–December 1998.

Project Management Institute, *The Standard for Portfolio Management,* 3rd edition (Newtown Square, PA: Project Management Institute, 2013).

Project Management Institute, Thought Leadership Series. *Implementing the Project Portfolio: A Vital C-Suite Focus,* November 2015.

Questions for Charles O'Reilly, Organizational Ambidexterity, *Rotman Magazine,* University of Toronto, Fall 2014.

Rumelt, Richard, *Good Strategy Bad Strategy: The Difference and Why it Matters* (Crown Business, July 19, 2011).

Stanier, Michael Bungay, *The Coaching Habit: Say Less, Ask More & Change the Way You Lead Forever* (Box of Crayons Press, February 29, 2016).

Sull, Donald, and Kathleen M. Eisenhardt, *Simple Rules: How to Thrive in a Complex World* (Houghton Mifflin Harcourt, April 21, 2015).

"Sunk Cost," Accessed November 7, 2017. https://en.wikipedia.org/wiki/Sunk_cost.

Weatherhead, P.J., "Do Savannah Sparrows Commit the Concorde Fallacy?" *Behavioral Ecology and Sociobiology.* Springer Berlin, 1979.

Weill, Peter, and Sinan Aral, "Managing the IT Portfolio: Returns from the Different IT Asset Classes," CISR, Sloan School of Management, MIT, March 2004. http://seeit.mit.edu/Publications/Weill-Aral-ITportfolio%2003-12-04.pdf.

"Yahoo Memo: The Peanut Butter Manifesto," November 18, 2006. https://www.wsj.com/articles/SB116379821933826657.

Chapter 5: Execution

Austin, Robert, and Lee Devin, "Artful Making: What Managers Need to Know About How Artists Work," *Financial Times,* May 8, 2003.

Bowles, Dennis L., and Darrel G. Hubbard, *A Compendium of PMO Case Studies: Reflecting Project Business Management Concepts* (PBM Concepts, 2012).

Brown, Tim, *Change by Design: How Design Thinking Transforms Organizations and Inspires Innovation* (Harper Business, September 29, 2009).

"Changepoint Predicts Top Technology Trends Facing Enterprise IT in 2017." http://www.prnewswire.com/news-releases/changepoint-predicts-top-technology-trends-facing-enterprise-it-in-2017-300382786.html.

Daft, Richard L. *Organization Theory and Design,* 10th edition (South-Western, CENGAGE Learning, 2010).

Duggal, Jack *Are You a Project Artist: The Skills of Project Artistry* (Proceedings of the Global North America PMI Congress, Washington DC, 2010).

Duggal, Jack *Managing the DANCE: Think Design, Not Plan* (Proceedings of the Global North America PMI Congress, Vancouver, Canada, 2012).

Duggal, Jack, *Managing the DANCE: The Pursuit of Next Generation PM Approach and Tools* (Proceedings of the EMEA PMI Congress, Milan, Italy, 2010).

Godin, Seth, *Linchpin: Are You Indispensable?* (Portfolio, April 26, 2011).

Ismail, Salim, Michael S. Malone, and Yuri van Geest, *Exponential Organizations: Why New Organizations Are Ten Times Better, Faster, and Cheaper Than Yours (and What to Do About It)* (Diversion Publishing, October 14, 2014).

Liedtka, Jeanne, and Tim Oglivie, *Designing for Growth: A Design Thinking Toolkit for Managers* (Columbia Business School Publishing, June 28, 2011).

Martin, Roger, *The Design of Business: Why Design Thinking Is the Next Competitive Advantage*, 3rd edition (Harvard Business Review Press, October 13, 2009).

Project Management Institute, *A Guide to the Project Management Body of Knowledge (PMBOK® guide)*, 6th edition (Newtown Square, PA: Project Management Institute, 2017).

Project Management Institute, Agile Certified Practitioner, Examination Content Outline, 2014.

Project Management Institute, *Implementing, Organizational Project Management (OPM), A Practice Guide* (Newtown Square, PA: Project Management Institute, March 2014).

Project Management Institute, *Rally the Talent to Win: Transforming Strategy into Reality*, PMI Thought Leadership Series, 2014.

Rinde, Sigurd, "Two Types of Workflow." http://blog.thingamy.com/.

Schmidt, Eric, and Jonathan Rosenberg, *How Google Works* (Grand Central Publishing, September 23, 2014).

Wideman, Max, "PM Glossary." Accessed June 10, 2017. http://maxwideman.com/pmglossary/index.htm.

Chapter 6: Governance

Barker, Eric, *Barking Up the Wrong Tree: The Surprising Science Behind Why Everything You Know About Success Is (Mostly) Wrong* (Harper One, May 16, 2017).

Cooper, Robert G., and Scott Edgett, "Best Practices in the Idea-to-Launch Process and Its Governance." *Research-Technology Management,* March 2012.

"Decision Rights Tools," Bain and Company Guide, January 10, 2015. http://www.bain.com/publications/articles/management-tools-decision-rights-tools.aspx.

Project Management Institute, *Governance of Portfolios, Programs, and Projects: A Practice Guide* (Newtown Square, PA: Project Management Institute, January 2016).

Project Management Institute, *Pulse of the Profession, The High Cost of Low Performance* (Newtown Square, PA: Project Management Institute, 2016).

"The Remarkable Self-Organization of Ants." Accessed July 31, 2017. https://www.theguardian.com/science/2014/apr/11/ants-self-organization-quanta.

Sull, Donald, and Kathleen M. Eisenhardt, *Simple Rules: How to Thrive in a Complex World* (Houghton Mifflin Harcourt, April 21, 2015).

Chapter 7: Connect

Ashkenas, Ron, "Jack Welch's Approach to Breaking Down Silos Still Works," *Harvard Business Review*, September 9, 2015.

Baker, Dave P., Rachel Day, and Eduardo Salas, "Teamwork as an Essential Component of High-Reliability Organizations," *Health Services Research*, Part II, August 2006.

Doyle, Andy "Management and Organization at Medium," March 4, 2016. https://blog.medium.com/management-and-organization-at-medium-2228cc9d93e9#.nedm3ja14.

Duggal, Jack, "Why You Need a Marketing Communications Plan for Your Project," Next Level Up column in *PMI Community Post*, September 12, 2011.

Finerty, Susan, *Master the Matrix: 7 Essentials for Getting Things Done in Complex Organizations* (Two Harbors Press, 2012).

Greve, Arent, Mario Benassi, and Arne Dag Sti, "Exploring the Contributions of Human and Social Capital to Productivity." Paper presented at SUNBEKT, XXVI, Vancouver, BC, April 25–30, 2006.

"Holocracy Self-Management Practice for Organizations." Accessed November 7, 2017. https://www.holacracy.org/.

Krebs, Valdis, "Social Network Analysis: An Introduction." http://www.orgnet.com/sna.html.

Laufer, Alexander, Edward J. Hoffman, Jeffrey S. Russell, and W. Scott Cameron, "What Successful Project Managers Do," *MIT Sloan Review Magazine*, March 16, 2015.

Moresco, Mariù and Carlo Notari, "Stakeholders' Worlds," in *Projects and Complexity*, edited by Francesco Varanini and Walter Ginevri (CRC Press, 2012).

Wenger, Etienne, *Communities of Practice: Learning, Meaning, and Identity* (Cambridge University Press, 1999).

"What Is Holacracy." Accessed October 25, 2017. https://www.zapposinsights.com/about/holacracy.

Chapter 8: Measure

Argyris, Chris, "Teaching Smart People How to Learn," *Harvard Business Review*, May–June 1991.

Ariely, Dan, *Payoff: The Hidden Logic that Shapes Our Motivations* (TED Books, Simon & Schuster/TED, November 15, 2016).

Barkai, Joe, *The Outcome Economy: How the Industrial Internet of Things Is Transforming Every Business* (Create Space Independent Publishing Platform, May 25, 2016).

Duggal, Jack, "How Do You Measure Project Success? Rethinking the Triple Constraint," Next Level Up column in *PMI Community Post,* July 9, 2010.

Duggal, Jack, "In the Pursuit of the Elusive: Showing PMO Value!" Paper presented at PMI® Global Congress—Asia Pacific, Kuala Lumpur, Malaysia (Newtown Square, PA: Project Management Institute, 2009).

Duggal, Jack, "Why Red May be Good for Your Project," Next Level Up column in *PMI Community Post,* October 9, 2009.

Grenny, Joseph, Kerry Patterson, David Maxfield, Ron McMillan, and Al Switzler, *Influencer: The New Science of Leading Change,* 2nd edition (McGraw-Hill Education, May 14, 2013).

Hammer, Michael, "7 Deadly Sins of Performance Measurement," *MIT Sloan Management Review,* Spring 2007.

Hassell, David, "How to Fit OKRs into Your Company's Mission and Values?" Accessed October 25, 2017. https://thenextweb.com/business/2017/03/08/fit-okrs-companys-mission-values/#.tnw_BhO2o86L.

Homem de Mello, Francisco S. "A Brief History of Objectives and Key Results," June 30, 2015. http://www.qulture.rocks/blog/2015/6/30/a-brief-history-of-objectives-and-key-results.

"Information Radiator." Accessed July 31, 2017. http://alistair.cockburn.us/Information+radiator.

"Performance Metrics and Measures." Accessed July 31, 2017. https://cio.gov/performance-metrics-and-measures/.

Poppendieck, Mary, and Tom Poppendieck, *Implementing Lean Software Development: From Concept to Cash* (Addison-Wesley Professional, September 17, 2006).

Projectize Group, LLC, USA, Project/Program Management Office (PMO) Survey (2005–2017).

Tufte, Edward, *The Visual Display of Quantitative Information* and *Envisioning Information* (Graphics Press, 2nd edition, January 2001).

"What Is Net Promoter Score." Accessed July 31, 2017. https://www.netpromoter.com/know/.

Chapter 9: Change

"100-culture-change-insights-from-100-culture-expert-posts. Accessed October 25, 2017. "http://www.cultureuniversity.com/100-culture-change-insights-from-100-culture-expert-posts/. "Understanding the Impact of Positive Deviance in Work Organizations," April 7, 2004. http://www.positivedeviance.org/pdf/publications/Understanding%20the%20Impact%20of%20Positive%20Deviance.pdf.

Boston Consulting Group, "DICE: A Tool for Executional Certainty." http://dice.bcg .com/index.html#intro.

Duggal, Jack, "Avoid Nagging, Make Routine Tasks a Habit," Next Level Up column in *PMI Community Post,* March 11, 2011.

Duggal, Jack, "How to Change the World: The Next Generation of Project Managers." Paper presented at PMI® Global Congress 2013, New Orleans, Louisiana (Newtown Square, PA: Project Management Institute).

Fothergill, Erin, et al. "Persistent Metabolic Adaption 6 Years after 'The Biggest Loser' Competition." *Obesity,* 2016.

Frank, Christopher J., and Paul F. Magnone, *Drinking from the Fire Hose: Making Smarter Decisions without Drowning in Information* (Portfolio, September 1, 2011).

Gawande, Atul, *The Checklist Manifesto: How to Get Things Right* (Metropolitan Books, December 22, 2009).

Grenny, Joseph, Kerry Patterson, David Maxfield, Ron McMillan, and Al Switzler, *Influencer: The New Science of Leading Change,* 2nd edition (McGraw-Hill Education, May 14, 2013).

Haidt, Jonathan, *The Happiness Hypothesis: Finding Modern Truth in Ancient Wisdom* (Basic Books, December 22, 2005).

HBO Films, *Temple Grandin.* http://www.hbo.com/movies/temple-grandin.

Heath, Chip, and Dan Heath, *Switch: How to Change Things When Change Is Hard* (Crown Business, February 16, 2010).

Hendricks, Benjamin, "What Is a Changemaker?" Blog post. Accessed October 25, 2017. http://changemakerobs.wordpress.com/what-is-a-changemaker/.

Kahneman, Daniel, *Thinking, Fast and Slow* (Farrar, Straus and Giroux, October 25, 2011).

Kegan, Robert, and Lisa Lahey, *Immunity to Change: How to Overcome It and Unlock the Potential in Yourself and Your Organization (Leadership for the Common Good)* (Harvard Business Review Press, January 13, 2009).

Kim, W. Chan, and Renée Mauborgne, *Blue Ocean Strategy: How to Create Uncontested Market Space and Make the Competition Irrelevant,* Expanded edition (Harvard Business Review Press, January 20, 2015).

Kotter, John P., *Leading Change,* 1R edition (Harvard Business Review Press, November 6, 2012).

Kubler-Ross, Elisabeth, *On Death and Dying* (Scribner Classics, July 2, 1997).

"Manifesto for Agile Software Development." 2001. Accessed May 19, 2017. agilemanifesto.org.

Montgomery, Sy, *Temple Grandin: How the Girl Who Loved Cows Embraced Autism and Changed the World* (HMH Books for Young Readers, April 3, 2012).

"Nudge Theory Toolkit." Accessed July 31, 2017. http://www.businessballs.com/ nudge-theory.htm#nudge-toolkit.

"Only One-Quarter of Employers Are Sustaining Gains from Change Management Initiatives, Towers Watson Survey Finds," Towers Watson, August 29, 2013. http://www .towerswatson.com/en/Press/2013/08/.

"Pestle Analysis." September 1, 2017, accessed October 7, 2017. https://www.cipd.co .uk/knowledge/strategy/organisational-development/pestle-analysis-factsheet.

"Prosci ADKAR® Model for Individual Change to Drive Organizational Transformation." Accessed October 7, 2017. https://www.prosci.com/adkar.

Project Management Institute, *Managing Change in Organizations: A Practice Guide* (Project Management Institute, 2013).

Projectize Group, LLC, USA, Project/Program Management Office (PMO) Survey (2005–2017).

Samson, Alain, *The Behavioral Economics Guide*, 2014. Accessed August 10, 2017. http:// www.behavioraleconomics.com.

Sirkin, Harold L., Perry Keenan, and Alan Jackson, *"The Hard Side of Change Management." Harvard Business Review*, October 2005.

Taleb, Nassim Nicholas, *The Black Swan: The Impact of the Highly Probable* (Random House, April 17, 2007).

Thaler, Richard H., and Cass R. Sunstein, *Nudge: Improving Decisions about Health, Wealth, and Happiness* (Yale University Press, April 8, 2008).

Thibaut, John, and Laurens Walker, *Procedural Justice: A Psychological Analysis* (Lawrence Erlbaum Associates, 1975).

Zachary, G. Pascal, "When Innovation Moves Too Fast," *IEEE Spectrum*. Posted June 19, 2017 | 20:00 GMT. http://spectrum.ieee.org/at-work/innovation/when-innovation-moves-too-fast.

Chapter 10: Learn

Argyris, Chris, "Teaching Smart People How to Learn (PDF). *Harvard Business Review*, May–June 1991.

Berger, Warren, *A More Beautiful Question: The Power of Inquiry to Spark Breakthrough Ideas* (Bloomsbury USA, 2014).

Bevelin, Peter, *Seeking Wisdom: From Darwin to Munger* (PCA Publications, 2007).

Bhargava, Rohit, *Non-Obvious 2017: How to Think Different, Curate Ideas and Predict the Future*, 2017 edition (Ideapress, December 6, 2016).

Christensen, Karen, "Amy Edmondson, Thought Leader Interview," *Rotman Magazine*, University of Toronto, Winter 2015.

Dobelli, Rolf, *The Art of the Good Life: 52 Surprising Shortcuts to Happiness, Success and Wealth* (Hachette Books, November 7, 2017).

Duggal, Jack, "Avoid Repeated Project Woes: Try a Different Approach to Lessons Learned," Next Level Up column in *PMI Community Post,* June 10, 2011.

Ferris, Timothy, *The 4-Hour Chef: The Simple Path to Cooking Like a Pro, Learning Anything, and Living the Good Life* (New Harvest, November 20, 2012).

Gawande, Atul, *The Checklist Manifesto: How to Get Things Right* (Metropolitan Books, December 22, 2009).

Gittleson, Kim, "Can a Company Live Forever?" *BBC News,* New York, January 19, 2012. http://www.bbc.com/news/business-16611040.

Harford, Tim, *Adapt: Why Success Always Starts with Failure* (Picador, 2012).

"How to Run a PANCAKE Retrospective. Accessed October 7, 2017." https://www.lynda.com/Business-Skills-tutorials/How-run-PANCAKE-retrospective/175961/468247-4.html.

Mahalakshmi, N., and Rajesh Padmashali, "Shane Parrish interview," *Outlook Business,* June 7, 2017. https://www.outlookbusiness.com/specials/a-weekend-in-omaha_2017/we-are-prone-to-overvaluing-complexity-in-reality-simplicity-makes-all-the-difference-3600.

"NASA Masters Forum: The Power of Storytelling. Accessed October 7, 2017." https://appel.nasa.gov/knowledge-sharing/masters-forums/.

"The Retrospective Starfish." March 9, 2006. Accessed October 7, 2017. http://www.thekua.com/rant/2006/03/the-retrospective-starfish/.

Ries, Eric *The Lean Startup: How Today's Entrepreneurs Use Continuous Innovation to Create Radically Successful Businesses* (Crown Business, 2011).

Scott, Susan, *Fierce Conversations: Achieving Success at Work and in Life, One Conversation at a Time* (Berkley, 2002).

Stanier, Michael Bungay, *The Coaching Habit: Say Less, Ask More & Change the Way You Lead Forever* (Box of Crayons Press, February 29, 2016).

Starmind. Accessed October 7, 2017. http://www.starmind.com/.

Syed, Matthew, *Black Box Thinking: Why Most People Never Learn from Their Mistakes—But Some Do* (Portfolio, 2015).

Taleb, Nassim Nicholas, *Antifragile: Things that Gain from Disorder,* Reprint edition (Random House Trade Paperbacks, January 28, 2014).

Tim Ferris Show, Podcast interview with Marc Andreesen. Accessed October 7, 2017. https://www.transcripts.io/transcripts/tim_ferriss_show/2016/05/28/marc-andreessen.html.

Chapter 11: Simplify

American Heritage Dictionary, 5th edition. (Houghton-Mifflin, 2012).

Chen, Brian, "Simplifying the Bull: How Picasso Helps to Teach Apple's Style Inside Apple's Internal Training Program," *New York Times,* August 10, 2014.

Duggal, Jack, "Kill the PMO, Build a Department of Simplicity." Proceedings of the Global North America PMI Congress, Phoenix, Arizona, 2014.

Emiliani, Bob, *Better Thinking, Better Results: Using the Power of Lean as a Total Business Solution* (The CLBM, LLC, January 1, 2003).

Gawande, Atul, *The Checklist Manifesto: How to Get Things Right* (Metropolitan Books, December 22, 2009).

"Hans Hoffman Quotes." Accessed October 25, 2017. http://www.hanshofmann.net/quotes.html#.WX6ZaumQxPY.

Jensen, Bill, Simpler work blog. Accessed October 25, 2017. http://www.simplerwork.com/.

Kaplan, Soren, *Leapfrogging: Harness the Power of Surprise for Business Breakthroughs* (Berrett-Koehler, 2012).

"Least Resistance: How Desire Paths Can Lead to Better Design." January 25, 2016, Accessed October 16, 2017. http://99percentinvisible.org/article/least-resistance-desire-paths-can-lead-better-design/.

Maeda, John, *The Laws of Simplicity (Simplicity: Design, Technology, Business, Life)* (MIT Press, 2006).

Morieux, Yves, "Smart Rules: Six Ways to Get People to Solve Problems without You," *Harvard Business Review,* September 2011.

Projectize Group, LLC, USA, Project/Program Management Office (PMO) Survey (2005–2017).

Roberts, Sam, "Dot-Dot-Dot, Dash-Dahs-Dash, No More," *New York Times,* February 12, 2006. http://www.nytimes.com/2006/02/12/weekinreview/dotdotdot-dashdashdash-no-more.html.

"Siegel + Gale, Global Simplicity Index." Accessed October 16, 2017. http://simplicityindex.com/.

Simplicity Advisory Board (SAB), Linda Tischler, "The Beauty of Simplicity," *Fast Company Magazine,* November 1, 2005.

Sull, Donald, and Kathleen M. Eisenhardt, *Simple Rules: How to Thrive in a Complex World* (Houghton Mifflin Harcourt, April 21, 2015).

Chapter 12: Balance

Burnam-Fink, Michael, interview, "Will Joel Garreau & Jamais Cascio Prevail—Along with the Rest of Us?" http://prevailproject.org/blog/2012/01/02/will-joel-garreau-jamais-cascio-prevail-%E2%80%94-along-with-the-rest-of-us/. January, 2, 2012, accessed November 30, 2017.

Gino, Francesco," Paradoxical Frames—The Benefits of Embracing Conflict," *Rotman Magazine,* University of Toronto, Winter 2013.

Heifetz, Ronald A., Marty Linsky, and Alexander Grashow, *The Practice of Adaptive Leadership: Tools and Tactics for Changing Your Organization and the World* (Harvard Business Press, May 18, 2009).

Laufer, Alexander, Edward J. Hoffman, Jeffrey S. Russell, and W. Scott Cameron, "What Successful Project Managers Do," *MIT Sloan Review Magazine,* Spring 2015.

Martin, Roger, The Opposable Mind: How Successful Leaders Win through Integrative Thinking, 1st edition (Harvard Business School Press, 2007).

Milan, Matthew, *The Next User You Design for Won't be Human, Fast Company,* Co.Design, November, 20, 2017 (https://www.fastcodesign.com/90146967/the-next-user-you-design-for-wont-be-a-human).

Rothenberg, Albert, *The Emerging Goddess: The Creative Process in Art, Science, and Other Fields* (University of Chicago Press, December 15, 1989).

Taleb, Nassim Nicholas, *Antifragile: Things that Gain from Disorder,* Reprint edition (Random House Trade Paperbacks, January 28, 2014).

Webb, Amy, "The Signals Are Talking: Why Today's Fringe Is Tomorrow's Mainstream," *MIT Sloan Review,* March 2017.

ACKNOWLEDGMENTS

Thanks to the thousands of participants of my seminars and workshops, and clients from organizations around the world who have engaged, questioned, challenged, experimented, and applied the ideas discussed in this book, and helped me fine-tune them over the last seventeen years. I am grateful for your stories and examples of transforming your organizations, creating magic with your teams, facing challenges and setbacks, winning awards, getting promoted, starting new endeavors and exploring what's next for you, that have inspired this book.

I am grateful to Margaret Cummins from Wiley for seeking me out after attending one of my talks, and supporting and standing by to get this project completed. Thanks to Jayalakshmi E T, Kalli Schultea, and the entire editorial and production team at Wiley.

Thanks to the many clients and practitioners—executives, PMO leaders and project, program and portfolio managers who have helped me to engage with them to *create theories from practice and experience,* rather than the other way around— Alina Grossman, Annika Andersson, Chris Richard, Colin Smith, Daniel Steeves, Elaine Barrera-Nicholson, Heather Pinto, Jayne Odom, Julie Allison, Julia DeSouza, Ken Jeans, Kurt Baraniecki, Mary Donnici, Mohamed Hammadi, Neeta Mhatre, Ray Woeller, Ricky Aladort, Shernette Barham, Stella Estorga, Steve Andersen, Veronica Thralls, Wayne Maddox, and many others who wish to remain anonymous. Appreciation for Sarina Arcari who applied the DNA of Strategy-Execution and won the PMO of the Year award and continues to evangelize and practice the ideas discussed in the book. Also, thanks to the Project Management Institute (PMI) and the PMI staff who I have had the opportunity to work with over the years.

Thanks to Dean Miller and Cory Sauls for sharing their practices in execution and governance as they continue to apply and transform their organizations. Also thanks to Bruce Harpham and Jeremy Gantz for lengthy discussions and assistance with a couple of chapters. Thanks to Beth Taylor for patiently reviewing parts of the manuscript and providing valuable feedback. Also, thanks to Beth and Leslie Maness for applying and evangelizing the ideas in different forums.

Special thanks to Laila Faridoon who has made a difference not just in her organization and her country, but is impacting her region and beyond by applying and living the principles of next-generation project management and leadership espoused in this book.

I also thank my friends and colleagues, who supplied hours of inspiration, feedback, ideas and encouragement: Brad Malone, Carl Pritchard, Dave Po Chedley, Gary Heerkins, Greg Githens, Harbeer Malhotra, James Patsalides, John Watson, Laxmi Parmeswar, Lisa Di Tullio, Markus Earhart, Maria Kolenda, Michel Thiry, Nick Shufro,

Regina Sggir, Sat and Sweety Duggal, Stig Villadsen, Suchitra Ramani, and Thomas Steinwender. Appreciation for Jonathan Gilbert for friendship and brotherhood as we began the journey of recognizing the DANCE and learning to surf on the edge of chaos a long time ago.

Also, eternal gratitude to Jai and Nina for being part of my DANCE and challenging me to deal with it better each day, while I was in the air for long stretches, juggling many things, while working on this book.

ABOUT THE AUTHOR

Jack Duggal is the founder and managing principal of Projectize Group LLC, leading next-generation strategy-execution, consulting, facilitation, learning and development, organizational assessments and transformation engagements. He works with leading organizations from NASA to Silicon Valley and governments around the world.

Jack is a change-maker, facilitator, coach, and transformational teacher with a passion to seek different perspectives in how to deal with the DANCE (Dynamic | Ambiguous | Non-Linear | Complex | Emergent) nature of business and organizations in today's turbulent world, and inspire people and organizations to realize their potential and make a difference, through transformational strategy execution.

An internationally recognized expert in Strategy-Execution and PMO, Jack's next-generation ideas and the DNA of Strategy-Execution framework have been adopted and implemented in many organizations around the world. He has worked with a broad range of executives, organizational team leaders and managers who have helped create theories from practice and experience, rather than the other way around. His group has built an impressive portfolio of clients and established a proven track-record with hands-on implementation of PMOs, design and facilitation of large scale-organizational transformation, and leadership and training programs with measurable results, in different industry verticals, over the last seventeen years.

Jack is a sought-after international keynote speaker and top-rated seminar leader and visiting faculty in the areas of Strategy-Execution, Agile, PMO, Program Management, Leadership, Design, Change, Connection, and Learning. He has worked with thousands of participants in seminars and workshops from organizations around the world who have engaged, questioned, challenged, experimented, applied, and helped fine-tune the practices over the years.

Jack lives on a hill in Connecticut on the east coast of United States with a vista that inspires a wide perspective, and runs on the trails in the valley to practice and gain experience in the trenches.

Connect with Jack Duggal at jduggal@projectize.com and visit www.projectize.com. Check dnaofstrategyexecution.com for additional resources.

INDEX

Note: Page references in *italics* refer to figures and tables.